# Electronics Primer:

## The Products, The Market, The Players

# Electronics Primer:

## The Products, The Market, The Players

Marlane Miller
Angela Moran
Brian Richardson
Tom Waguespack
Russell Carter
Warren Bickley

Edited by, M. Laura Kane-Sellers

**CorePurpose Publishing, Inc.**
**Phoenix, Arizona**

First Edition



Kane-Sellers, M. Laura, 2005
Electronics Primer: The Products, The Market, The Players

ISBN 978-0-9747056-8-2

Manufactured in the United States of America

# Acknowledgments

Grateful acknowledgment is made to the following Texas A&M University students for their participation in the IDIS 420 classroom project that culminated as this work.

Marlena Anderson
Buck Bailey
Warren Bickley
Tom Blankmeyer
Luke Boyea
Will Bridenstine
Richard Brown
Adam Burger
Bill Carroll
Russell Carter
Ernie Cates
Katie Christian
Jennifer Christian
Derek Clark
Justin Clark
Adam Curley
Heath Davis
Craig Denard
Gary Dennis
Will Dillard
Sean Dugan
Jeff Dunn
Ben Dutton
Jonathon Edwards
David Entringer
Lucas Evans
Kristen Fisher
Jacob Flamm
Zach Fogle
Clint Ford
Susan Foster
Benton Freeman
Brad Gaultney
Jackie Gohlke
Jessica Guerra
Tim Guidry
Andy Havel
Kellie Hayman
Julie Hemann
Michael Henderson
Zac Henderson
Scott Hill
Jerry Hoefler
Deanna Howard
Brent James
Mitch Jeter
Andrew Jones
Cory Junemann
Tom Keiley
Brian Kelly
Jon Kleckley
Kevin Kolos
Nate Korce
Kristin Koury
Cal Leavitt
Travis Ledford
Micah Lightfoot
Jason Lilly
Justin Lopez
Lacee Lynch
Jeff Martin
Josh Mason
Taylor McGaughey
Robert McGehee
Brian Meitz
Marlane Miller
Marisa Montague
Justin Moore
Angela Moran
Brad Morris
Neal Mukherjee
Shahin Naghavi
Travis Nash
Adeena Neves
Jared New
Benjamin Newberry
Lynda Nguyen
Michael Nguyen
J.J. Novak
Jed Nowlin
Michael Nucholls
Tanya Ochoa
Dennis Olesen
Mario Ortega
Daniel Pattillo
Brinnet Paul

John Peterson
Josh Poole
Justin Pouk
Steven Potter
Andy Ramirez
Matthew Ranker
John Rehus
Allison Reynolds
Matt Richards
Brian Richardson
Brett Ridgeway
Aaron Roy
Jack Russo
David Saalfeld
Jason Sanchez
Matt Sandoval
Ryan Savage
Paul Schaller
Erik Schindewolf
Deanna Schulte
Jordan Schumacher
Tye Scott
Ryan Skaggs
Stephen Skrabanek
Joseph Sodolak
Wiratama Somiadiredja
Jeff Spencer
Colin Sproule
Ann Stavinoha
Noah Stavlo
Eric Stolpman
James Sullivan
Justin Tankersley
Ross Thigpen
Nate Thomas
Austin Thompson
Katie Thurman
Chris Todd
Quinn Tolbert
Dustin Truitt
Tori Vaughan
Eric Weaver
Megan Wilke
Brandon Winn
Justin Wolf
Sarah Yesavage
Nathan Zatopek

Additionally we would like to thank the following individuals for their assistance and support:

Robin Gray, NEDA
Dr. F. Barry Lawrence, Texas A&M University, ID Program
Joan Koerber-Walker, CorePurpose, Inc.
Mark Taylor, Taylor Creative Marketing
John Ward, iSuppli
Doug Andry, SIA
John Simari, Texas Instruments, Inc.
Kevin Sink, TTI
Frank Robertazzi, CML

The editor takes this opportunity to thank the student authors – Marlane, Angela, Brian, Tom, Russell, and Warren. You did a great job and I hope that you are as proud of your accomplishment as I am. Best of luck in your careers and future endeavors. It has been a true pleasure working with each of you on this project and taking a small part in your professional development.

M. Laura Kane-Sellers

Publication of this book was made possible through a grant from the NEDA Education Foundation, which is affiliated with the National Electronic Distributors Association (NEDA). The authors and editor wish to express our gratitude for this generous gift and endorsement.

**National Electronic Distributors Association**

The National Electronic Distributors Association (NEDA) is the leading not-for-profit trade association representing supplier authorized distributors of electronic components and their manufacturer partners. Dedicated to enhancing the distributor/manufacturer relationship, NEDA represents companies involved in the distribution of electronic components, computer and computer peripheral components and test, measurement and control equipment parts. Members of the organization consist of supplier authorized distributors and manufacturers of electronic components and parts, computer and computer peripheral components, and test, measurement, and control equipment.

As an active advocate for the authorized electronics distribution industry, NEDA advances its value proposition through research, marketing and the one-of-a-kind Supply Chain Savings Calculator. Through white papers and round table forums, NEDA is instrumental in integrating industry needs to drive standards on critical issues. NEDA maintains an active presence with the leading universities of Industrial and Electronic Distribution degree programs in an effort to prepare the students of today to be the leaders of tomorrow. To assist members with gaining and maintaining a competitive edge in the dynamic business environment, NEDA provides access to superior professional educational programs that are tailored to address the challenges inherent in the electronics distribution industry.

NEDA has exclusive rights for the sale and distribution of this book. To order your copy or for bulk orders please visit ***www.nedassoc.org*** and select Publications or call the headquarters office at 678-393-9990.

National Electronic Distributors Association
1111 Alderman Drive, Suite 400, Alpharetta, GA 30005-4175
PH: 678-393-9990 FAX: 678-393-9998 ***www.nedassoc.org***

The authors and editor also extend great appreciation and recognition to iSuppli Corporation and the Semiconductor Industry Association (SIA) for their important contributions and continued support.

iSuppli Corporation – the global leader in electronics industry research – provides market intelligence services for the contract manufacturing, original equipment manufacturing (OEM), and supplier communities.

World-class, multinational companies utilize iSuppli databases, tools and market analysis reporting to improve decision making when it comes to what products to source, which partners to select, what capital investments to make; and where in the world to perform each of these endeavors.

iSuppli research is tailored to meet the needs of the supply chain, with frequent updates that stress immediate issues like pricing, lead times, and availability. iSuppli offers diverse suites of services that enable users and suppliers of materials, electronic components, and equipment to improve performance in specific areas:

Increase demand,
More efficiently allocate capacity,
More effectively manage inventories and production plans,
Enhance targeting and timeliness of new product introductions,
Improve delivery and customer service,
Benchmark or measure performance,
Reduce supply chain cycle times,
Reduce overall supply chain risks, disruptions, and liabilities, and
Lower costs.

More information is available at ***www.isuppli.com***.

---

The Semiconductor Industry Association (SIA) is the premier trade association representing the U.S. semiconductor industry. Founded in 1977 by five microelectronics innovators, SIA unites 95 companies responsible for more than 85% of semiconductor production in the U.S.

The coalition provides domestic semiconductor companies a forum to advance the global competitiveness of the $80 billion U.S. chip industry. Through a network of corporate CEOs and working committees, SIA shapes public policy on issues critical to the industry and provides a spectrum of services to aid members in growing their own business.

More information about the Semiconductor Industry Association can be found at ***www.sia-online.org.***

# About the Editor

**M. Laura Kane-Sellers (Laurie)** is a full-time faculty member at Texas A&M University in the Industrial Distribution Program. In addition to her responsibilities at TAMU, Laurie serves as Education Foundation Director for the National Electronic Distributors Association (NEDA). She worked in the electronics industry for 20 years prior to pursuing her Ph.D. at TAMU. Having completed all Ph.D. coursework in Education Human Resource Development with a minor in Industrial Distribution, she hopes to complete her dissertation by December 2005. Laurie also holds two master's degrees: one in Business from Southern Methodist University, and the other in Speech Communications from Texas Tech University. Laurie is co-author of *Connecting Resources: A Primer for Electronics Distribution* published in 2004. She is currently working on a book titled *Painting the Homeless Visible and Giving them a Voice* to be released in late 2005.

Pictured: (Back Row Left to Right) Marlane Miller, Brian Richardson, Laurie Kane-Sellers, Russell Carter; (Front Row Left to Right) Warren Bickley, Angela Moran, Tom Waguespack.

# About the Authors

**Marlane Miller** is a Senior Industrial Distribution student at Texas A&M University, class of 2005. She will graduate in December of 2005 with a bachelor's of science degree in Industrial Distribution along with a minor in Business. Marlane is at present seeking a job in the electronics industry with plans to obtain an MBA in the future.

**Angela Moran** is from Dallas, Texas, and a graduate of the Texas A&M Class of 2004. She majored in Industrial Distribution with a minor in Mathematics. Currently, Angela is observing in the Dallas Independent School District to complete her secondary teacher certification. After this, she plans to obtain a master's degree.

**Brian Richardson** was born and raised in San Antonio, Texas. Brian graduates in December of 2005 with a bachelor's of science degree in Industrial Distribution from Texas A&M University. He is currently searching for a job in the electronics industry and looks forward to marrying his high school sweetheart in March of 2006.

**Tom Waguespack** is a May 2005 graduate of the Texas A&M University Industrial Distribution Program. He is currently an Account Manager for TTI, Inc., an industry leading distributor of passive, interconnect, and electromechanical components. Tom was born and raised in Houston, Texas.

**Russell Carter** grew up in Houston, Texas, and attended Texas A&M University studying Industrial Distribution. Russell found his niche in the electronics field and was inducted into the Sigma Delta honor society. He graduated in May 2005 and accepted a job with Texas Instruments in Dallas, Texas, where he is responsible for capacity/business planning for worldwide manufacturing of analog products.

**Warren Bickley** graduated from Texas A&M University in May of 2005 from the Industrial Distribution Program. Warren lives in Dallas, Texas and, along with Tom, is an Account Manager for TTI, Inc.

# Contents

# Semiconductors (Cont'd)



# Interconnect, Passive & Electromechanical

# Introduction

The consolidation in the electronics distribution industry over the last two decades has enabled distributors that began as storefront operations on Radio Row in the early 1900s to grow into multi-billion dollar, publicly traded global companies. As their revenue and customer reach expanded, so did their line card of authorized suppliers, and the number of products occupying the shelves of their primary distribution centers around the world. This intensive strategic approach places tremendous demands on the distribution sales force. Customers expect their sales representatives to be knowledgeable about the products they sell. For distributor sales people, getting a grasp on hundreds of thousands of products is a daunting if not impossible task.

*Electronics Primer: The Products, The Market, The Players* is intended to provide a desk reference for distributor sales people regardless of whether they are just entering this dynamic industry as a new hire or have been working in distribution for years. As distribution firms have grown and become more sophisticated in technical acumen and logistic processes, the complexity of developing and maintaining a well-trained sales force requires a significant investment in human capital resources. As Education Foundation Director for the National Electronic Distributors Association (NEDA), I receive frequent inquiries regarding the availability of tools that can sharpen the skills of professional sales personnel. Until now, there has not been a comprehensive, all-inclusive handbook or desk reference that provides a basic explanation of the products and their features, the applications and customers, the market size and future trends, or the manufacturers and their authorized distributors.

This reference book was born from a classroom project in the Electronics Distribution Networks course, part of the world-renowned Industrial Distribution (ID) program at Texas A&M University (TAMU). Each chapter of the book began as a student team project, and then was refined and formatted by the authors—all senior level students within the ID program. Intrigued by the electronics industry, three of the six authors have graduated and accepted jobs with a leading electronics distributor or manufacturer. Throughout the book, you will find referenced material that manufacturers, distributors, and market research firms provided on a complimentary basis because they recognized the market need that this book fulfills. The authors and editor are extremely grateful to these firms. We acknowledge them for both their assistance and the confidence they placed in us to fill a knowledge gap that we believe this book closes.

Each chapter of the book follows a similar format, beginning with a basic definition of the product–hence the book's partial title *The Product.* In defining the product, various technologies used in the manufacturing process are discussed. Unlike technical manuals or manufacturers' data books, this primer provides a plain English explanation of the product and its corresponding attributes. Readers progress through each chapter learning first about the product type, its product features and user benefits, and end-product applications. This allows distributor salespeople to be better prepared to target customer opportunities within their assigned territories. Ideally, we would have taken an end-product, block-diagram approach, but as electronic components have proliferated the marketplace, this would have required adopting a generic approach that ultimately would yield minimal benefit in terms of specificity.

The features/user benefits portion of *The Product* section translates key product attributes into customer care-abouts and end-equipment application considerations. Missing from most manufacturers' date sheets is the translation of features into benefits in an easy-to-understand format. For a distributor sales person to represent and promote products adequately, it is important that he or she recognizes the key attributes that customers use in making product specification decisions. This section of the book arms salespeople with the knowledge and vernacular pertinent to the various product types on their supplier authorized line card.

The second subsection of each chapter is titled *The Market.* In this section, we identify the worldwide total available market (TAM), market segmentation by region of the world as well as targeted customer market segment, and revenue growth trends and projections. Painstaking efforts have been taken to verify and validate market data. Through a technique called triangulation or multiple source checks, the TAM, segmentation, and growth trends have been developed for each chapter and are contained in the sections titled Market Trends and Market Segmentation. Triangulation involves the development of the initial conclusion, then checking multiple experts to validate the information. This process was facilitated by several leading market research sources for subscription data, namely World Semiconductor Trade Statistics (WSTS), from the Semiconductor Industry Associations, and Applied Market Intelligence provided by iSuppli, as well as various publicly available sources and media articles. Every effort has been made to ensure the accuracy of this information. The authors and editor welcome inputs and corrections for future editions of this primer.

The final section of each chapter, *The Players,* directs readers to the manufacturers of the focus product in that chapter. Generally accepted sources have been used to develop the ranked lists of manufacturers/suppliers. The websites and manufacturer reviews of each chapter facilitated development of the Key Players matrix found in each of the chapters. While it is impossible to list all of the suppliers and authorized distributors for every product, we have attempted to list the major players. (Again, if you would like to submit inputs for the next edition of this book, we welcome your feedback.)

*Electronics Primer: The Products, The Market, The Players* also contains a glossary of commonly used terms. Throughout the book, the reader will encounter terms in boldface. That indicates that these terms are defined in the glossary of the book. Additionally, we have provided an index to facilitate quick reference of specific material contained within this book.

Introduction

It is serendipitous when a classroom project can be transformed into a tool that serves a market need. As editor of the work, I am especially proud of the work that these students have done. (The students who participated in the initial projects are identified and acknowledged in the beginning of the book.) The authors worked diligently to refine these projects and produce the book that you are holding. Throughout the project, I have developed a special relationship with the six young adults named in the About the Authors section. I would like to take this opportunity to express my gratitude and pride in the outstanding work that they have created.

As the consumer of this product, if you have feedback, modifications, or ideas to be considered for the next edition, please submit your inputs to ***lks75248@aol.com***. Thank you for purchasing this book. We hope it will be a useful tool in your career in the electronics distribution industry.

Your purchase of this book is an investment in the future. All royalty proceeds are being donated into a scholarship fund for Industrial Distribution students.

M. Laura Kane-Sellers

# Semiconductors

# 1

# Active Discretes

## Diodes, Thyristors, Transistors

The term **discrete** refers to a packaged semiconductor device having a single device. Discretes are devices that have an electrical functional component such as a diode, thyristor, or transistor. These devices, unlike integrated circuits (**IC**s), must be combined with other components to provide a basic electrical function such as amplification, switching, or latching. Discrete devices preceded the introduction of digital integrated circuits, and are considered a mature product family. Yet, discrete semiconductors continue to grow in numbers of suppliers, by application and by technology[12].

The active discrete devices can be categorized into three primary groups[14]:

- Uncontrolled: **Diode**
- Semi-controlled: **Thyristor—**SCR
- Fully controlled: **Power transistors**—BJT, MOSFET, IGBT, GTO, IGCT

**Figure 1:** Discrete Device Hierarchy[12]

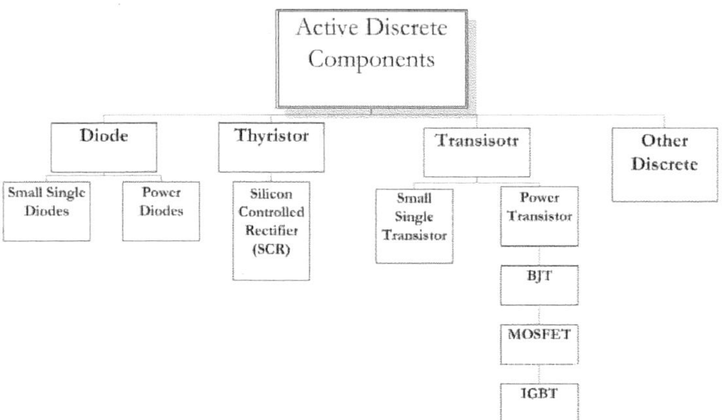


A **diode** is a simple semiconductor device that allows current to flow in one direction but not the other. It is basically the electronic version of a one-way valve. It consists of a positive/negative (PN) junction and has two terminals, an anode (+, P-type) and a cathode (-, N-Type), and current flows from anode to cathode within the diode[2].

**Figure 2:** Standard Diode Symbol[3]

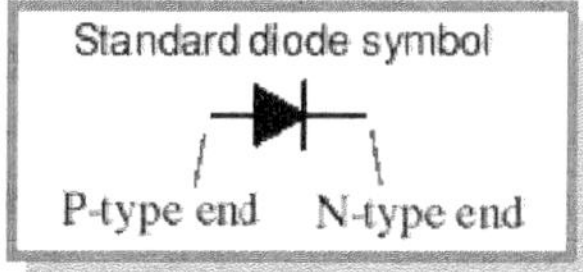


The amount of current flow through the diode depends on the direction of the applied voltage. If voltage is applied from the cathode to anode (reverse biased or blocking state), only a small leakage current flows through the diode until the reverse breakdown voltage is reached. But when voltage is applied the other way, anode to cathode (forward biased), it conducts current with a small forward voltage across it[14].

A simple example of a diode function is a light switch in a home or building. When the switch is closed, the circuit is completed and the light turns on. When the switch is open, there is no current and the light is off. However, a semiconductor diode's behavior is not ideal[3]. When voltage is in the reverse biased direction, there is a small leakage of current until the breakdown voltage is reached and the junction breaks down letting current through. Usually, the breakdown voltage is a lot more voltage than the circuit will ever see, so it is irrelevant. When forward biased, a small amount of voltage is needed to start the hole-electron combination process at the junction and get the diode going, thus allowing current to flow freely[1] (see Figure 3). This is because the actual current/voltage relationship depends on the detail of how the diode was made—the choice of materials, doping, etc. [3]

**Figure 3:** Current vs. Voltage Graph[3]

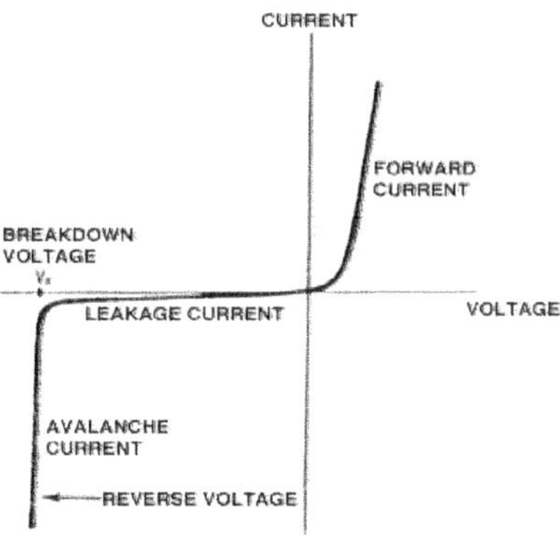


Germanium, Selenium, Gallium Arsenide and Silicon are what diodes are made from; however, most diodes are made of Silicon [4]. There are many different types of diodes made; the most common types are listed below in Figure 4.

**Figure 4:** Types of Diodes[4]

| Diode Type | Definition | Symbol |
|---|---|---|
| **Rectifier Diode** | A heavy-duty diode that uses high current to convert AC to DC [4]. | Rectifier Diode |
| **Tunnel Diode** | A diode with a negative resistance region. The current through the diode increases up to a specific point and then decreases to a specific point, and the cycle begins again[4]. | Tunnel Diode |
| **Zener Diode** | A diode designed to conduct in the reverse [bias] direction, which means it will not let current flow until it has reached the break down voltage[23]. | Zener Diode |
| **Varactor Diode** | Also called a varicap diode or tuning diode. It is used in electronic circuits primarily as a voltage-controlled capacitor, and its diode function is secondary. It is operated reverse-biased so no current flows through it[19]. | Varactor Diode |
| **Schottky Diode** | A semiconductor diode with a low forward voltage drop. The typical application is a discharge-protection for solar cells connected to lead-acid batteries. Standard silicon diodes have a forward voltage drop of 0.6 volts, compared to the Schottky's 0.2 volts[16]. | [6] |
| **Light Emitting Diode (LED)** | An LED is a semiconductor device that emits incoherent monochromatic light when electrically biased in the forward direction. The color depends on the semiconducting material used, and can be near-ultraviolet, visible, or infrared[9]. | LED |
| **Photodiode** | An electronic component and a type of photodetector that is a p-n junction designed to be responsive to optical input. It is provided with either a window or optical fiber connection in order to let in the light to the sensitive part[19]. | Photo Diode |
| **Varistor** | A type of resistor with a non-ohmic current-voltage characteristic. Varistors are commonly used to protect circuits against excessive voltage by acting as a spark gap[20]. | Varistor |

The **thyristor** is a solid-state semiconductor device similar to a diode that has an extra terminal called a **gate**, which is used to turn it on. It acts much like a switch that latches on. It is a four-layer semiconductor device, consisting of alternating P type and N type materials (PNPN), usually having three electrodes: an anode, a cathode, and a gate[18]. Each device is turned on at the appropriate time by a trigger pulse applied to the gate and the device will remain on until the load current through it drops to zero [5]. When switched on, it can only pass current in one direction (like a diode)[19].

Thyristors are mainly used where high currents and voltages are involved, and are often used to control alternating currents whereby modifications of alternating currents cause the device to automatically switch off[19].

The most common type of thyristor is the **silicon-controlled rectifier (SCR)**. Just like a diode, in the normal "off" state, the device restricts current flow to the leakage current and, when the gate to source current exceeds a certain point, the device turns "on" and conducts current. The SCR will remain in the "on" state even after gate current is removed as long as current through the device remains above the holding current (essentially the device is latched in the "on" state). Once current falls below the holding current, the device will switch off. If the applied voltage increases rapidly, it may induce enough leakage current to trigger the device into the on state, but this can be harmful to the device[19].

SCRs are used in power switching, phase control, chopper, battery chargers, and inverter circuits. Using this type of thyristor, large amounts of power can be switched or controlled using a small triggering current or voltage. In industry, they are applied to produce variable DC voltages for motors from AC line voltage. They can also be used in some electric vehicles to modulate the working voltage. Another common application is phase control circuits used with inductive loads. SCRs can also be found in welding power supplies. There, they are used to maintain a constant output current or voltage[19].

Modern thyristors can switch large amounts of power. In the realm of very high power applications, they are still the primary choice. However, in low and medium power applications, they have almost been replaced by other devices with superior switching characteristics like MOSFETs or IGBTs[19].

One major problem associated with the thyristor is that it is not a fully controllable switch in the sense that triggering current direction needs to be reversed to switch it off. **GTO (Gate Turn-off Thyristor)** is another related device that addresses this problem[19].

Thyristors come in many other types as well, such as:

- Mosfet Controlled Thyristor (MCT): Two additional FET structures for on and off control.
- Mosfet Turn On (MTO).
- Insulated Gate Commutated Thyristor (IGCT).

- Static Induction Thyristor (SITh) or Field Controlled Thyristor (FCTh): A gate structure can shut down anode current flow
- Triacs: Dual polarity thyristors[14].
- Diacs.
- Phase controlled: Rectifying line frequency voltage and current, for AC and DC motor drives, large voltage and current capability, low on-state voltage drop[14].
- Inverter grade: Used in inverter and chopper, fast, can be turned-on using forced commutation[14].
- Light activated: Similar to phase controlled, but triggered by pulse of light. Normally very high power ratings[14].

The transistor, invented by three scientists at the Bell Laboratories in 1947, rapidly replaced the vacuum tube as an electronic signal regulator[18]. Prior to the invention of transistors, digital circuits were composed of vacuum tubes, which had many disadvantages. They were much larger, required more energy, dissipated more heat, and were more prone to failures. It's safe to say that without the invention of transistors, computing as we know it today would not be possible[21].

A **transistor** is a small electrical device found in everything from radios to robots. They have two key functions: (1) they can amplify an electrical signal (modulator) and (2) they can switch on and off, letting current through or blocking it as necessary[21]. In the simplest sense, the transistor works like the dimmer in a room; push the knob of the dimmer, the light comes on; push it again, the light goes out—a *switch*. Rotate the knob back and forth, and the light grows brighter or dimmer—a *modulator*. To understand amplification, think of this: A relatively effortless action by you to turn the knob from its low to high setting translates into a much greater a reaction by the light—the whole room is lit up—an *amplifier*[22].

**Figure 5:** Symbol of the Transistor[22]

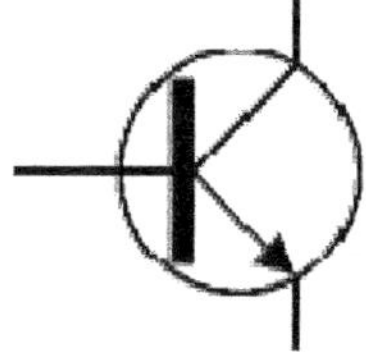

A transistor has the ability to control its own semiconductance, namely acting as a conductor when needed or as an insulator when needed. A transistor consists of three layers of a semiconductor material, each capable of carrying a current. The transistor's three-layer structure contains an N-type semiconductor layer sandwiched between P-type layers (a PNP configuration) or a P-type layer between N-type layers (an NPN configuration). Putting a thin piece of semiconductor of one type between two slices of another type has startling results: A little current in the central slice is able to control the flow of the current between the other two. That little current in the middle is supplied by an antenna or another transistor, for example. Even when the input current is weak—as from a radio signal that's traveled a great distance—the transistor can control a strong current from another circuit through itself. In effect, the current through the "output side" of the transistor mimics the behavior of the current through the "input side." The result is a strong, amplified version of the weak radio signal[22].

**Types of Transistors [4]:**

- **BJT**: Bias Resistor Transistors, or Digital Transistors—small signal bipolar NPN or PNP transistors equipped with integrated biasing and pull-down resistors
- **IGBT** : Integrated Gate Bipolar Transistor.
- **FET**: Field Effect Transistor
- **JFET**: Junction FET (Field Effect Transistor)
- **MOSFET**: Metal Oxide Semiconductor Field Effect Transistor
- **MESFET**: Metal Semiconductor Field Effect Transistor (FET).
- **GaAs**: Gallium Arsenide.
- **HEXFET, TMOS, DMOS, UMOS, VMOS, ZMOS**: Additional names that define the internal structure, not type.

# The Product

## Applications and End Uses

Diodes are used in many different ways. For example, a device that uses batteries often contains a diode that protects the device if you insert the batteries backward. The diode simply blocks any current from leaving the battery if it is reversed, which protects the sensitive electronics in the device[1].

Other applications of diodes include[19]:

- Radio demodulation of amplitude modulated (AM) radio.
- Logic gates: Diodes can be used to construct logic gates.
- Power conversion: Half wave rectifier, diode bridge or full wave rectifier, double diodes, alternators.
    - **Half wave rectifier**—a diode used to convert alternating current electricity into direct current by removing the negative portion of the current
    - **Diode Bridge/ single-phase bridge rectifier / full wave rectifier**—a special arrangement of four diodes that will transform an alternating current into a direct current,

using both positive and negative excursions of a single phase alternating current.
    - **Three phase full wave rectifier**—only six diodes needed.
    - **Double diodes**—a pair of diodes in the same housing; with a split (center-tapped) alternating current supply, full wave can be obtained.
    - **Alternators**—devices that generate three-phase alternating current
- Over-Voltage Protection

Thyristors are used in many applications including: motor speed controls, pressure-control systems, and liquid-level regulators[18], variable-speed electric motors, power supplies for electrochemical processes, lighting and heating control, and controllers for electric utility power systems[17].

Thyristor units are particularly cost-effective for fast systems, for complex loads involving transformers, and/or heaters whose resistance changes with temperature or time. They are the best means of controlling electrical heating power. Thyristor units generally offer improved controllability, which gives economy and easy serviceability—particularly with the plug-in versions that reduce down time and maintenance costs. Almost silent in operation, the units can eliminate the switch on surge by operating in synchronization with the supply, and therefore offer an improved working environment[5].

Transistors are the basic elements in integrated circuits. In a microchip, a very large number of transistors are interconnected in a particular pattern or design so that the amplified output of one drives another[22]. Manipulating the pattern of transistors can create many different functions such as[22]:

- Timers to control watches or ovens.
- Sensors to monitor temperatures, detect intruders, or control car wheels from locking.
- Logic or arithmetic processors to perform computations in calculators and computers, process words, search, and run systems to transmit voice, data, graphics, and video.

Transistors are very durable, very small, and have a high resistance to physical shock. They are also inexpensive. Although discrete transistors are still used, the vast majority of transistors are built as parts of integrated circuits. Transistors are used in virtually all electronic devices, including radio receivers, computers, space vehicles, and guided missiles[8].

The active discrete category includes diodes, thyristors, transistors, diodes, and other discretes. These devices, unlike integrated circuits, must be combined with other components to provide a basic electrical function such as power conversion, motor control, amplification, switching, or latching.

## Product Features and User Benefits

**Figure 6:** Basic Product Features and User Benefits of Diodes

| Product Features | User Benefits |
|---|---|
| Versatile Packaging | Can be manufactured in different packages and different materials to comply with different applications. |
| Adaptable to Different Applications | Diodes can be used as voltage regulators, in switching applications, in RF circuits, or to make logic decisions in digital circuits. |
| Accepts Extreme Voltages and Currents | Diodes are created for different applications and can be made to withstand very high voltages and currents, as well as very low ones. |

**Figure 7:** Product Features and User Benefits of Thyristors

| Product Features | User Benefits |
|---|---|
| Adjustable to different Applications | Thyristors are used as switches, rectifiers, or voltage regulators. |
| Can be Latched | Once a thyristor is turned on, it remains on. |
| Can conduct in one or two directions | Some thyristors conduct like a diode, only in one direction. Others can conduct in both directions. |
| Variable State Control | State of the thyristor can be controlled in different ways, such as light or terminals. |

**Figure 8:** Product Features and User Benefits of Transistors

| Product Features | User Benefits |
|---|---|
| Amplify Signals | Can intensify a very small current to a large current. |
| Ability to Control Large Currents | Available to control hundreds of amperes. (A vacuum tube that controls even one ampere is large and costly.) |
| No warm up time | Do not require any time to warm up or prepare before use. |
| Low Cost | As opposed to its predecessor, the vacuum tube, can be very low cost in high volumes. |

## Life Cycle

**Figure 9:** Life Cycle Curve for Selected Active Discretes

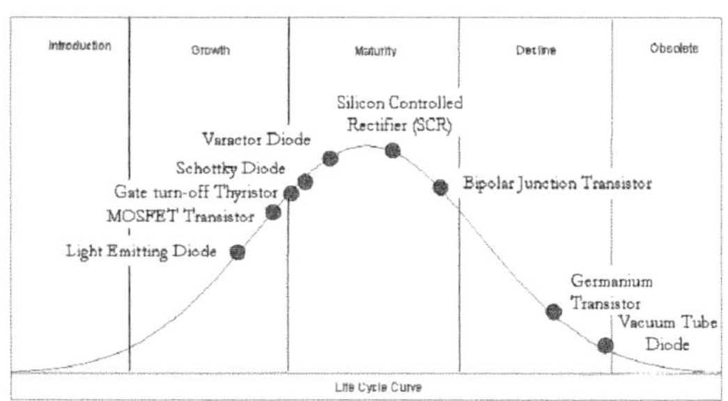


# The Market

## Total Available Market

**Figure 10:** Revenue History and Forecast of Worldwide Semiconductors and Discrete Production ($M)[7]

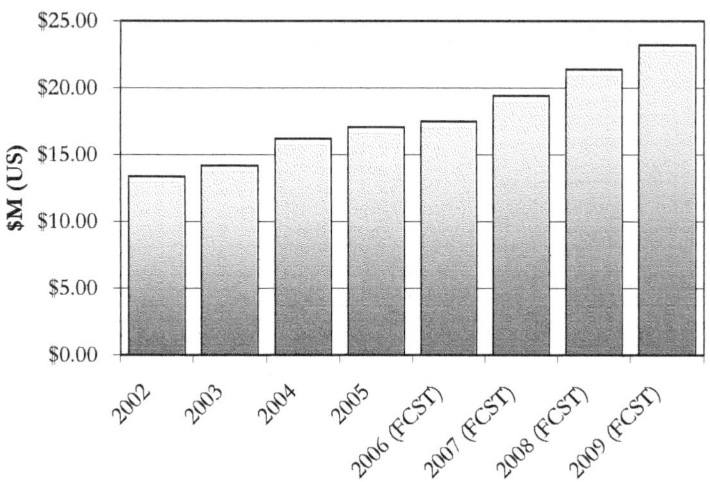


| 2002 | 2003 | 2004 | 2005 | 2006 (FCST) | 2007 (FCST) | 2008 (FCST) | 2009 (FCST) |
|---|---|---|---|---|---|---|---|
| $13.4 | $14.2 | $16.2 | $17.1 | $17.5 | $19.4 | $21.4 | $23.2 |

## Market Segmentation

**Figure 11:** Discrete Market Segmentation

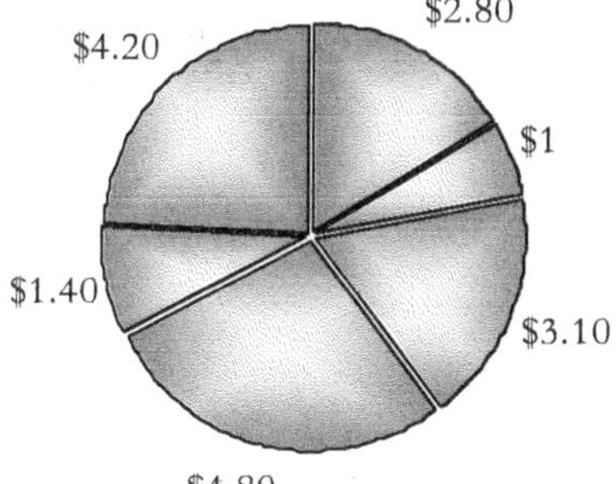


| Discrete Market Segmentation for 2005 | |
|---|---|
| Data Processing | $2.80 |
| Wired Communications | $1.00 |
| Mobile Communications | $3.10 |
| Consumer Electronics | $4.80 |
| Automotive Electronics | $1.40 |
| Industrial Electronics | $4.20 |

Source: iSuppli Corporation

## Regional Segmentation

**Figure 13:** Worldwide Discrete Revenue by Regional Producer ($M US)[13]

| | 2000 | 2001 | 2002 | 2003 |
|---|---|---|---|---|
| American Companies | $3.57 | $3.74 | $3.78 | $3.98 |
| Japanese Companies | $5.23 | $6.39 | $6.59 | $6.88 |
| European Companies | $3.57 | $2.44 | $2.64 | $2.82 |
| Asia/Pacific Companies | $5.53 | $0.33 | $0.39 | $0.53 |
| **Total** | $17.90 | $12.90 | $13.40 | $14.20 |

# The Players

## Key Players

**Figure 14:** Top 20 Suppliers of Total Discrete, 2004 $B (US)[11]

| | Supplier | Website | Distributors | Revenue $B (US) | Market Share % |
|---|---|---|---|---|---|
| 1 | Toshiba | www.toshiba.com | Arrow; Avnet; NuHorizons; TAEC | $1.30 | 8% |
| 2 | ST Microelectronics | www.st.com | Arrow; Avnet; Digi-Key; Future; Mouser; NuHorizons | $1.22 | 7% |
| 3 | Renesas Technology | www.renesas.com | All American; Arrow; Avnet; Digi-Key; NuHorizons | $1.16 | 7% |
| 4 | Rohm Electronics | www.rohm.com | Digi-Key; Future | $1.06 | 6% |
| 5 | Fairchild Semiconductor | www.fairchildsemi.com | All American; Arrow; Avnet; Future | $0.97 | 6% |

| | Supplier | Website | Distributors | Revenue | Market |
|---|---|---|---|---|---|
| 6 | Philips Semiconductor | www.philips.com | Arrow; Avnet; Digi-Key; Future | $1.30 | 6% |
| 7 | Infineon Technologies | www.infineon.com | Arrow; Avnet; Digi-Key; Future; Richardson | $1.22 | 6% |
| 8 | Vishay (General Semiconductor) | www.gensemi.com | Allied; Arrow; Avnet; Future; Garrett; Jaco; Mouser; Newark InOne; Passive Components; PUI; | $1.16 | 6% |
| 9 | International Rectifier (IR) | www.irf.com | ACI; Allied; Arrow; Avnet; C&H; Digi-Key; Future; Galco Industrial; Newark InOne; | $1.06 | 5% |
| 10 | Mitsubishi | www.mitsubishichips.com | All American; Arrow; Diamond-AC; PS2 Solutions | $0.97 | 4% |
| | | **Total** | | **$15.85** | |

## References


1 Brian, M. (n.d.). How Stuff Works. Retrieved Jun. 02, 2005, from How Semiconductors Work: Creating a Diode website: <http://electronics.howstuffworks.com/diode3.htm>.

2 Diode I/V Behavior. (n.d.). Retrieved Jun. 02, 2005, from University of St. Andrews. website: <www.st-andrews.ac.uk/~www_pa/Scots_Guide/info/comp/passive/diode/chars/chars.htm>.

3 Diode—Rectifier Tutorial. (n.d.). Retrieved Jun. 02, 2005, from American Microsemiconductor. website: <http://www.americanmicrosemi.com/tutorials/diode.htm>.

4 Davis, L. (n.d.). Leroy's engineering website. Retrieved Jun. 02, 2005, from Diode Manufacturers website: <http://www.interfacebus.com/diode.html>.

5 Eurotherm, (n.d.). What is a thyristor?. Retrieved Jun. 02, 2005, from <http://www.eurotherm.co.uk/uk/eng/Knowledge/PowerControl/What.htm>.

6 Georgia State University, (n.d.). Varactor. Retrieved Jun. 02, 2005, from <http://hyperphysics.phy-astr.gsu.edu/hbase/electronic/varactor.html>.

7 Gordon, Richard. (2005). 2Q05 Semiconductor Forecast Update: Fears of a Recession in 2005 Diminish (G00127762), Retrieved Jul 05, 2005, from Gartner database.

8 Invention and Uses of the Transistor. (2005). Columbia Encyclopedia, Sixth Edition. Retrieved Jun. 02, 2005, from Encyclopedia.com website: <http://www.encyclopedia.com/html/section/transist_InventionandUsesoftheTransistor.asp>.

9 LED. (2005). Retrieved Jun. 02, 2005, from wikipedia website: <http://en.wikipedia.org/wiki/LED>.

10 Metal Oxide Varistor. (2005). Retrieved Jun. 02, 2005, from wikipedia website: <http://en.wikipedia.org/wiki/Varistor.>.

11 Olsson, M. (2005). International Rectifier's Analog and Discrete Markets Continue to Grow (G00127657), Retrieved Jul 05, 2005, from Gartner Database.

12 Olsson, M. (1999). The Power Behind Discretes (SEMI-WW-DP-9909), Retrieved Jun 02, 2005, from Gartner database.

13 Olsson, M. (2003). Worldwide Semiconductor Market Trends: Discretes, 2003 (SEMC-WW-EX-0279), Retrieved Jun 02, 2005, from Gartner database.

14 Salam, Dr. Z. (2003). Power electronics and drives. 3-2003. Retrieved Jun 02, 2005, from: <http://ncon.fke.utm.my/courses/see_5433/intro.pdf>.

15 Schottky diode. (2005). Retrieved Jun. 02, 2005, from wikipedia website: <http://en.wikipedia.org/wiki/Schottky_diode>.

16 Semiconductor component manufacturers list. (2004). Retrieved Jun. 02, 2005, from Interfacebus website: <http://www.pc-control.co.uk/transistor.htm>.

17 Thyristor. (1997). Retrieved Jun. 02, 2005 from Infoplease website: <http://www.infoplease.com/dictionary/thyristor>.

18 Thyristor. (2004). Retrieved Jun. 02, 2005, from Tech Target site: <http://whatis.techtarget.com/definition/0%2C%2Csid9_gci778571%2C00.html>

19 Varacter diode. (2005). Retrieved Jun. 02,2005, from wikipedia website: <http://en.wikipedia.org/wiki/Varactor_diode>.

20 Varistor. (2005). Retrieved Jun. 02, 2005, from wikipedia website: <http://en.wikipedia.org/wiki/Varistor>.

21 What is a transistor? (2004). Retrieved Jun. 02, 2005, from <http://webopedia.internet.com/TERM/t/transistor.html>.

22 What is a transistor and how does it operate?. (2004). Retrieved Jun. 02, 2005, from PC Control website: <http://www.pc-control.co.uk/transistor.htm>.

23 Zener diode. (2005). Retrieved Jun. 02, 2005, from wikipedia website: <http://en.wikipedia.org/wiki?Zener diode>.

# Analog

# 2
# Analog Interface

**Interface** is any point, area, or surface along which two substances meet. More specifically in electronics, interface is a device or IC that sends or receives signals from other parts, ICs, or subsystems within the application[6]. Interface devices are used to integrate and connect the analog and digital elements of a system. These circuits are used to bridge the gap between the voltages of an analog device that often operate at much higher voltages. Since interface devices are used primarily in conjunction with digital products, they can reflect the volatility of demand inherent in the digital segment of the market[12]. Basically interface is a part or a circuit in a subsystem that sends or receives signals to or from other subsystems.

Analog Interface encompasses a wide array of products. Many devices overlap into other product areas but the primary products/topics covered in this chapter include:

- Line Drivers
- Receivers
- Transmitters
- Transceivers
- Modulators/Demodulators
- Universal Serial Bus (USB)
- Codec
- Recommended Standards

A **line driver** is a device that extends the transmission distance between terminals and computers. Also called a "short-haul modem" or "limited-distance modem," line drivers can extend a signal that is normally limited to a range of a few hundred feet to several miles[1]. It enhances the reliability, or consistency, of transmission over larger distances by driving the input to the transmission line with a higher than normal signal level[7].

A **(data interface) receiver** is a device that accepts signals and decodes it as opposed to a transmitter which sends signals. [11]. The term is used generically to refer to "the side being sent to," or to whatever device or **peripheral** happens to be at the other end of the communication[1]. A receiver is the recipient (observer) of a message (information), which is sent from a source (object).

A **(data interface) transmitter** is a device that generates and amplifies a **carrier wave, modulates** it with a meaningful signal derived from speech or other sources, and radiates the resulting signal from an antenna[15]. A transmitter, sometimes abbreviated **XMTR**, sends signals and is an object (source) that sends information to an observer (receiver)[16].

A **(data interface) transceiver** is a device within one chassis that performs both transmitting and receiving functions in a common housing. It is sometimes designed for portable or mobile use. Transceivers use common circuit components for both transmitting and receiving, which provides half-duplex operation[14].

**Modulators** are devices that perform modulations (modifications); modulation describes a range of techniques for encoding information on a carrier wave[8]. A **demodulator** performs the opposite function and decodes. It is an electronic circuit used to recover the information content from the carrier wave of a signal. The term is usually used in connection with radio receivers, but there are many kinds of demodulators used in many other systems. A common use is in a **modem**, which is a contraction of the terms modulator/demodulator[2].

**Universal Serial Bus (USB)** is an emerging standard for interconnecting PCs with peripherals. But it does more than just communicate digital information; it is also a power source. USB eliminates the need for an external AC power source by carrying sufficient power to make the deployment of computer peripherals easier than ever before[10]. USB provides a serial, bidirectional 12Mbit/second interface. It combines a multitude of existing interfaces into a single easy-to-use connector, greatly reducing system complexity and offering manufacturers the ability to develop highly integrated products[17].

The USB standard was initially drafted by a consortium of companies, including Intel, DEC, Microsoft, IBM, Compaq, NEC, and Northern Telecom. The primary goals of USB are to address users' needs for simplicity as well as provide multimedia capabilities. USB addresses these issues by dramatically simplifying the use of the PC with its "**Plug and Play**" feature and standard environment. This will enable the PC to emerge as true consumer product[17].

A **codec** is a device that encodes or decodes a signal through compression and decompression techniques. It combines analog to digital conversion and visa versa (digital to analog) in a single chip. A/D (analog to digital) and D/A (digital to analog) will be discussed in the data converter chapter. The codec uses an algorithmic technique to reduce the number of bytes required by large files and programs. This compression capability positions the codec ideally for use in modem products.

Electronic data communications between elements fall into two broad categories: single-ended and differential. Standards have been developed to ensure compatibility between units provided by different manufacturers, and to allow for reasonable success in transferring data over specified distances and/or data rates[13]. The Electronics Industry Association (EIA) has produced standards for **RS485, RS422, RS232,** and **RS423** that deal with data communications.

**Figure 1:** Data Communications Recommended Standards Hierarchy[13]

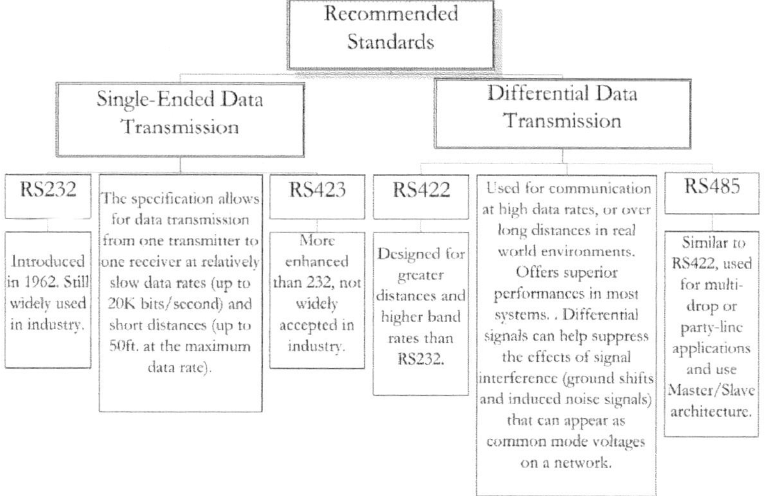


# The Product

## Applications and End Uses

Line drivers are used to connect POS terminals, machine tools, sensors, and countless other digital devices to a host computer at speeds typically ranging up to 128 Kbps[1]. An example of a line driver is an amplifier used to extend the range of an RS-232C digital signal beyond 50 feet while maintaining a specified bit-error ratio[7].

USB circuits are used in the Plug and Play products that we use to interface with computers to play music, view videos, and synchronize handheld devices with our desktop and notebook computers[17].

Codecs are typically used in modem products because of their compression capability.

## Product Features and User Benefits

**Figure 2:** Product Features and User Benefits of Analog Interface Products

| Product | Features | | User Benefits |
|---|---|---|---|
| **Codec** | Efficient Power Usage | ⇨ | Reduces the need for costly power supplies. |
| | High Accuracy | ⇨ | Sends signals with little error to enable reliable and consistent transmission/conversion of data. |
| | Number of Outputs | ⇨ | Enables flexibility, and stretches its ability to take on larger transmission/conversion loads. |
| | Adjustable Data Rates | ⇨ | Increases the range of usage for different users depending on the amount of speed required. Contains the ability to transfer information faster to speed up your system. |
| **Line Drivers** | Drives Data at Fast Rates | ⇨ | Transfers information faster (up to 19,200 bps), at longer distances, to speed up your system. |
| | Compact and Plug-and-Play Capabilities | ⇨ | Enables products with smaller size and ability to plug right in to the computer and eliminate the need for a stand-alone unit. Requires no separate outlet or power supply. |
| | Asynchronous Transmission | ⇨ | Transmission scheme adapts better to real world signals that are not always uniform in nature. This allows for data to be transmitted in a "non-parallel" manner. |
| | Excellent Noise Rejection | ⇨ | Increases the reliability of the signal transfers by filtering out superfluous noise caused by speed and signal distortion. |
| **Receivers** | Sensitivity | ⇨ | Ability to detect signals according to product application requirements. |
| | Selectivity | ⇨ | Will reject unwanted signals creating a faster, smoother receivability. |

| | | | |
|---|---|---|---|
| | Low Current Consumption | ⇨ | Power efficiency reduces the need for expensive power supplies. |
| **Transmitters** | High Accuracy and Linearity | ⇨ | Sends signals with little error to enable reliable and consistent transmission of data. |
| | Low Current Consumption | ⇨ | Power efficiency reduces the need for expensive power supplies. |
| | Universal | ⇨ | Designed to interface with any computer. |
| | Selectivity | ⇨ | Will reject unwanted signals creating a faster, smoother receivability. |
| **Transceivers** | Low Power Consumption | ⇨ | Reduces the need for costly power supplies. |
| | Thermal Shutdown and Short Circuit Protection | ⇨ | Protection from overheating and power surges that can damage the system. |
| | Limited **Slew Rates** | ⇨ | Ensures consistency of the signal, improving data integrity. Decreases the number of errors that may occur. |
| **Modulator Demodulator** | Excellent Carrier Suppression | ⇨ | Permits higher power levels in the side bands that are not usually possible in conventional **Amplitude Modulation (AM)** transmission. |
| | Adjustable Gain and Signal Handling | ⇨ | Provides flexibility for specific signal transmission application. |
| | Balanced Inputs and Outputs | ⇨ | Minimizes noise and interference that may enter system externally. At long lengths, balanced connections offer excellent induced noise rejection. |
| **USB** | Plug-and-Play | ⇨ | Immediately compatible with your system without the use of any external wiring or other hardware. Simplifies interface and reduces set-up time. |

| | | | |
|---|---|---|---|
| | Standardized Plug | ⇨ | Eliminates extra ports for more compact, cleaner design as well as saving money on extra gear. Also provides flexibility and ease of use. |
| | Hot Swapping | ⇨ | Allows user to plug in peripherals when the system is in use. Occasionally a device may require system restart, but not a complete power-down. |
| | Its Own Power Source | ⇨ | Eliminates bulky and expensive power supply units. |
| **RS Differential and Single-Ended Drivers** | **Fail-Safe** Receiver Inputs | ⇨ | Allows the elimination of cable fail-safe resistors, reducing power consumption. |
| | Smaller (232) | ⇨ | Smaller size promotes space-saving capabilities, cost reduction, and efficiency. |
| | High Data Rates (485) | ⇨ | Much faster transfer rates within the computer (or peripheral) that will save time and increase efficiency. |
| | Limited Slew Rate for Low EMI (482) | ⇨ | Ensures consistency of the signal, improving data integrity. Decreases the number of errors that may occur. |
| | Multiple Drivers and Receivers on the Same Bus (485) | ⇨ | Enables flexibility and stretches its ability to take on larger transmission loads. |

## Life Cycle

As mentioned earlier, RS 232 was the first recommended standard to be developed in 1962. RS 423 was developed later as an upgrade to RS 232. RS 423 was not widely accepted by the industry and therefore has slipped down the life-cycle curve, as RS 232 is still one of the most widely used transmission lines in the electronics and telecommunications industry. RS 422, developed for larger distances, has taken a stance in the industry while its successor, RS 485, is still relatively young, yet beginning to grow into the industry[13].

Typically, analog interface components adhere to the recommended standards. Therefore, the recommended standard supported by an analog product must be utilized to verify where that product lies on the life-cycle curve. Figure 3 depicts the life-cycle curve of the recommended standards.

**Figure 3**: Life Cycle Curve for RS (Differential and Single-Ended)

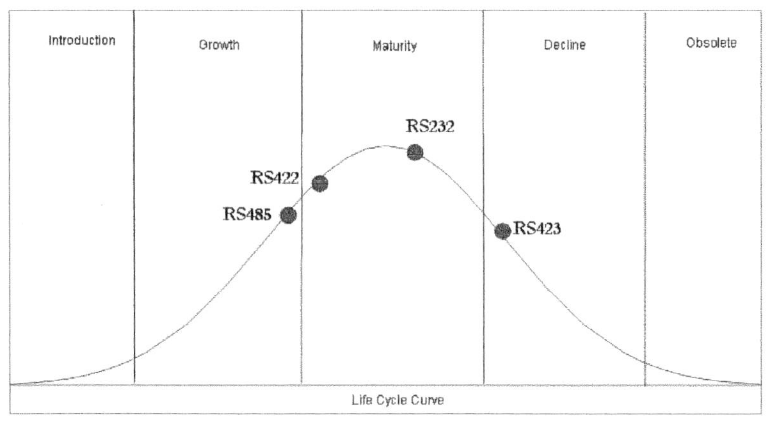


# The Market

## Market Trends

**Figure 4**: Analog Interface TAM and Market Trends [4;5]

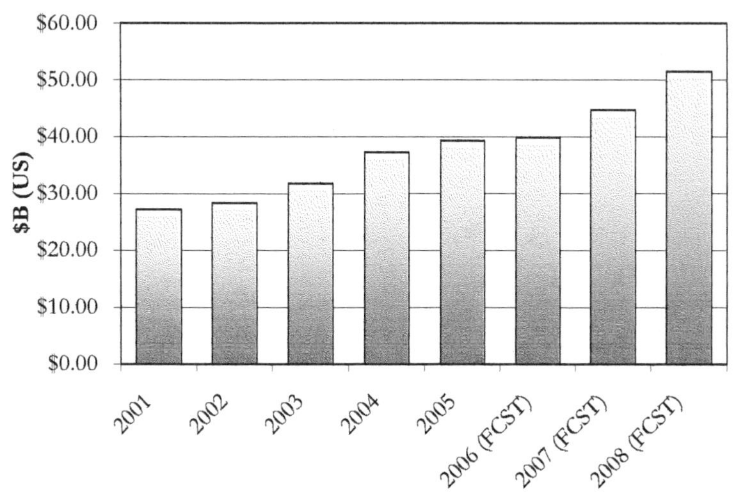


Source: iSuppli Corporation

| 2001 | 2002 | 2003 | 2004 | 2005 | 2006 (FCST) | 2007 (FCST) | 2008 (FCST) |
|---|---|---|---|---|---|---|---|
| $27.2 | $28.3 | $31.8 | $37.3 | $39.3 | $39.9 | $44.7 | $51.5 |

## Market Segmentation

**Figure 5:** Analog Interface 2004 Application Market Segmentation[12]

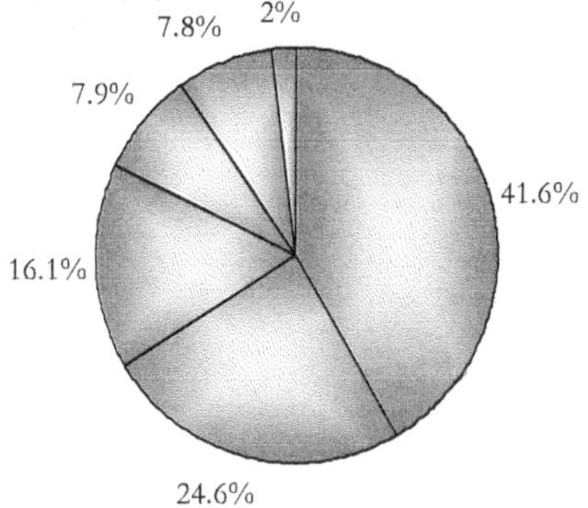


| Analog Interface Market Segmentation by Application | |
|---|---|
| Data Processing | 41.6% |
| Communications | 24.6% |
| Consumer Electronics | 16.1% |
| Industrial | 7.9% |
| Automotive | 7.8% |
| Military/ Aerospace | 2.0% |

# The Players

## Key Players

**Figure 6**: Analog Interface Key Player Summary[9]

| | Supplier | Website | Distributors | Revenue $M (US) | Market Share % |
|---|---|---|---|---|---|
| 1 | Texas Instruments (TI) | www.ti.com | Arrow; Avnet; Digi-Key; Newark InOne | $800 | 52% |
| 2 | National Semiconductor (NSC) | www.national.com | Arrow; Avnet; Digi-key; Future | $170 | 11% |
| 3 | International Rectifier (IR) | www.irf.com | ACI; Arrow; Avnet; Future; Universal | $116 | 8% |
| 4 | Maxim Integrated Products | www.maxim-ic.com | Arrow; Avnet; Digi-Key; Jaco; Newark InOne; Dallas Direct (Maxim) | $93 | 6% |
| 5 | Toshiba | www.toshiba.com | Arrow; Avnet; NuHorizons; TAEC | $82 | 5% |

| | Supplier | Website | Distributors | Revenue $M (US) | Market Share % |
|---|---|---|---|---|---|
| 6 | Linear Technology | www.linear.com | Arrow; Chip Supply; Digi-Key; Linear Express; Minco; NuHorizons | $69 | 4% |
| 7 | Rohm Electronics | www.rohmelectronics.com | Digi-Key; Future; | $36 | 2% |
| 8 | Sipex | www.sipex.com | Future | $33 | 2% |
| 9 | Analog Devices | www.analog.com | Arrow; Avnet; Digi-Key; Rochester; Semi Dice | $31 | 2% |
| 10 | Intersil | www.intersil.com | Arrow; Avnet; Chip Supply; Digi-Key; Minco; Newark InOne; Rochester | $23 | 1% |

## References


1 Computer Desktop Encyclopedia. (2005). Retrieved Mar. 21, 2005 from Computer Language Company, Inc. website: <http://www.computerlanguage.com/techweb.html>.

2 Demodulator. (2005). Retrieved Jul. 14, 2005 from Wikipedia website: <http://en.wikipedia.org/wiki/Demodulator>.

3 Floyd. T.J. (2003). *Digital Fundamentals*, 8ed. Pearson Education, Inc.

4 Henkel, R. (2001). How new system interfaces will generate chip revenue. Retrieved Apr. 7, 2005 from EETimes website: <http://www.eetimes.com/news/semi/showArticle.jhtml?articleID=10808640>.

5 Inouye, S. (2003). Interface. Retrieved Apr. 7, 2005 from Electronics Publications website: <http://www.electronics.ca/reports/ic/interface.html>.

6 Interface. (2005). Retrieved Jul. 14, 2005 from Wikipedia website: <http://en.wikipedia.org/wiki/interface>.

7 Line Driver. (2005). Retrieved Jul. 14, 2005 from Wikipedia website: <http://en.wikipedia.org/wiki/Line_driver>.

8 Modulation. (2005). Retrieved Jul. 14, 2005 from Wikipedia website: <http://en.wikipedia.org/wiki/Modulator>.

9 Olsson, M. (2005). Analog Market Reveune Increase in 2004. Retrieved Apr. 8, 2005 from Gartner database.

10 Universal Serial Bus (USB) Products. (2005). Retrieved Mar. 12, 2005 from Micrel, Inc. website: <http://www.micrel.com>.

11 Receiver. (2005). Retrieved Jul. 14, 2005 from Wikipedia website: <http://en.wikipedia.org/wiki/Receiver>.

12 Reilly, N. (2004). General-Purpose Analog/Mixed-Signal Landscape, 2004. Retrieved Mar. 17, 2005 from Gartner database.

13 Smith, R. (2005). RS485 Quick Reference. Retrieved Mar. 28, 2005 from RESmith website: <http://www.rs485.com/rs485spec.html>.

14 Transceiver. (2005). Retrieved Jul. 14, 2005 from Wikipedia website: <http://en.wikipedia.org/wiki/Transceiver>.

15 Transmitter. (2005). Retrieved Jul. 14, 2005 from Answers website: <http://www.answers.com/topic/transmitter?method=6>.

16 Transmitter. (2005). Retrieved Jul. 14, 2005 from Wikipedia website: <http://en.wikipedia.org/wiki/Transmitter>.

17 Universal Serial Bus (USB). (2005). Retrieved Jul. 14, 2005 from Wikipedia website: <http://en.wikipedia.org/wiki/USB>.

# 3

# Power Management

**Power Management Integrated Circuits or ICs** are small devices that regulate the power supplied to another device. These devices provide better efficiency and flexibility to everyday products such as cell phones and computers[5]. Power Management deals with regulating and controlling the amount of power used. It accomplishes this by regulating voltage and using converters to change the kind of current used in a device, such as a cell phone.

**Integrated Circuits (ICs)** are thin chips consisting of at least two interconnected semiconductor devices, primarily transistors. These chips also have passive components, like resistors, as well. The main product groups that are involved with Power Management ICs are Linear Voltage Regulators, AC-DC Converters, and DC-DC Converters[2].

There are several types of Power Management ICs including:

- Battery monitors
- Battery chargers
- Charge pumps
- Voltage references
- Voltage suppressors

This chapter will focus on the most common Power Management IC components: Linear Voltage Regulators, AC-DC Converters, DC-DC Converters, and Voltage Supervisors.

**Linear Voltage Regulators** are used to provide constant DC output voltage. They contain circuitry that continuously holds the output voltage at the design value regardless of changes in load current or input voltage. The linear voltage regulator is the basic building block of nearly every power supply used in electronics. A linear regulator operates power by using a voltage-controlled current source to force a fixed (set) voltage to be sent to the regulator output. The control circuitry in the regulator monitors the output voltage, and adjusts the current to hold the output voltage at the desired value. When a change in output voltage is detected, a control element, such as a transistor, is used to compensate, maintaining a constant voltage. Voltage regulators are used in power supplies that hold their output voltage constant over a wide range of load variations[3].

**AC-DC Converters** accept AC input voltage from a wall outlet and send DC voltage. The two configurations for the converters are wall mount and desk mount. In a wall mount setting, the adapter plugs directly into a wall, while in a desk mount setting the adapter is in-line between the AC plug and the DC output. Many DC devices use this type of adapter, or power supply, such as household items, printers, and low-voltage DC boards. Common features of the adapters include **over-voltage** protection, over-current protection, and short circuit protection.

**DC-DC Converters** accept DC input and provide regulated and/or isolated DC output in various applications, including computer flash memory, telecommunications equipment, process control systems, and vehicle-mounted systems. DC-DC Converters allow products to be smaller. Some of the common features for DC-DC Converters include constant current supply, **over-voltage** protection, **over-current** protection, short circuit connection, and remote on and off[1].

A voltage supervisor provides the maximum flexibility for monitoring and overlooking power supplies, as well as battery-controlled functions in systems, without the use of backup batteries. A **Voltage Supervisor** is a protection device that monitors certain system limitations and boundaries. It reacts to protect the system when those parameters go out of control (tolerance). The reset circuit (Voltage Monitor) function monitors the main system supply voltage and resets the system in case the voltage is too high or too low. Supervisors also protect against **runaway software**, high temperatures, and intrusion or system interference[4].

# The Product

## Applications and End Uses

Some of the applications for Power Management ICs include CD-ROM drives, cell phones, desktop chargers, desktop computers, digital cameras, mp3 players, and personal digital assistants (PDAs). These devices integrate numerous functions in a single electronic device, and Power Management ICs are used to regulate the multiple voltage output functions that come with these devices.

## Product Features and User Benefits

The features of Liner Voltage Regulator, AC-DC Converters and DC-DC Converters are very similar. Power Management Integrated Circuits focus on the amount of power that is flowing through the circuit. These features prevent destruction of the devices involved and allow for flexible use; therefore, almost any brand of these products will have these features. The usage and features of Power Management Integrated Circuits have remained the same since conception. As technology continues to advance, however, improvements will be made to increase performance and efficiency for Power Management ICs. Manufacturers continue to focus on size-reduction of analog integrated circuits.

**Figure 1:** Product Features and User Benefits of Linear Voltage Regulators

| Product Features | | User Benefits |
|---|---|---|
| Voltage-Controlled Sources | ⇨ | Forces a fixed voltage to appear at the output terminal. This allows you to use any small device without worry that there will be a voltage overload. |
| Dropout Voltage | ⇨ | A set maximum voltage that can be reached before the regulator drops out. This is extra protection that the system will not receive a voltage surge that can damage or destroy it. |
| Reverse Voltage Protection | ⇨ | Safeguards the system from shifts in voltage direction, a condition that can damage or destroy the system. |
| **Over-Current** Protection | ⇨ | Current output is shut down or limited in a situation from over-current or excessive charge. This will prevent any problems, such as the destruction to the regulators. |
| Can Have More Than One Output | ⇨ | Circuit can accommodate multiple currents simultaneously, reducing the need for additional devices. This saves power and board real estate as well as leads to better reliability due to fewer integrated circuits in the system. |
| Error Flags | ⇨ | Error flags provide useful information to the designer regarding operability. |

**Figure 2:** Product Features and User Benefits of AC-DC Converters

| Product Features | | User Benefits |
|---|---|---|
| Converts AC to DC | ⇨ | Allows you to use your DC devices with the AC current wall connection. This converter makes using DC devices possible. |
| Plugs Into a Wall | ⇨ | Provides the ability to plug the converter into the wall–increases convenience while reducing the cost. |
| Wide-Range of Voltage Alternating Current (VAC) | ⇨ | Ability to alternate current provides flexibility and reduces number of analog circuits required. This feature enables products to be used throughout the world regardless of a specific country's current standards. |
| Can Have More Than One Output | ⇨ | Allows you to use the converter on multiple DC devices if needed. Reduces the number of ICs required. |
| Shape, Size, and Configuration Vary | ⇨ | Design flexibility for specific application limitations. |
| **Over-Voltage** Protection | ⇨ | Voltage output is shut down or limited in a situation of **over-voltage**. This will prevent any problems such as the destruction of the DC device and data loss at the output end of the converter. |
| **Over-Current** Protection | ⇨ | Current output is shut down or limited in a situation from over-current. This will prevent any problems such as the destruction to the regulators. |
| Short Circuit Protection | ⇨ | Will increase the chances of data recovery if there is a short circuit on the load. If there is a short circuit, you have a better chance of not losing your information that was stored in your DC device. |

**Figure 3:** Product Features and User Benefits of DC-DC Converters

| Product Features | | User Benefits |
|---|---|---|
| Converts DC to DC | ⇨ | Regulates or isolates DC output in various DC devices. This allows for smaller devices to be used because the amount of DC can be limited to what is required for a small application. |

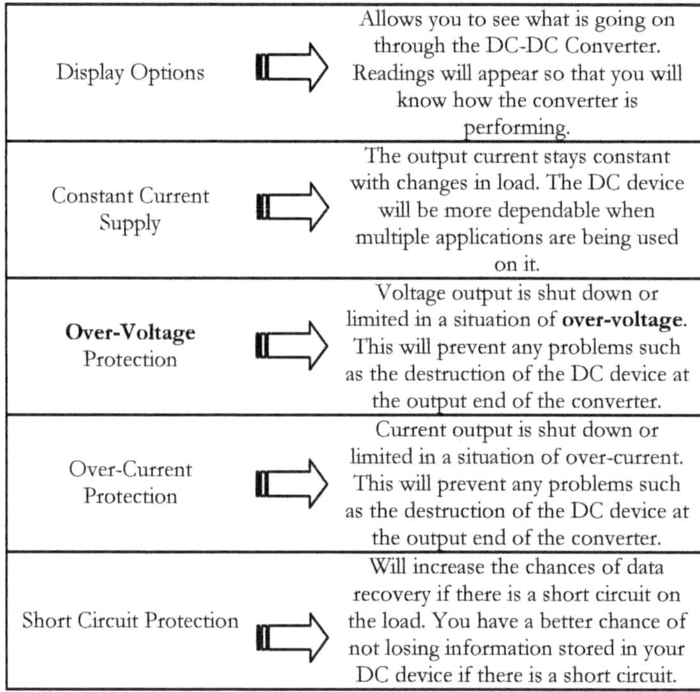

| | |
|---|---|
| Display Options | Allows you to see what is going on through the DC-DC Converter. Readings will appear so that you will know how the converter is performing. |
| Constant Current Supply | The output current stays constant with changes in load. The DC device will be more dependable when multiple applications are being used on it. |
| **Over-Voltage** Protection | Voltage output is shut down or limited in a situation of **over-voltage**. This will prevent any problems such as the destruction of the DC device at the output end of the converter. |
| Over-Current Protection | Current output is shut down or limited in a situation of over-current. This will prevent any problems such as the destruction of the DC device at the output end of the converter. |
| Short Circuit Protection | Will increase the chances of data recovery if there is a short circuit on the load. You have a better chance of not losing information stored in your DC device if there is a short circuit. |

**Figure 4:** Product Features and User Benefits of Voltage Supervisors

| Product Features | User Benefits |
|---|---|
| Precision Supply-Voltage Monitor | Protects the microprocessor (and the system) by monitoring the system power (Vcc). Sends a RESET signal to the µP (microprocessor) if the system power is out of tolerance and protects against unpredictable behavior. This will prevent problems, such as the destruction of your system. |
| Active-High and Active-Low Reset Outputs | Provides versatile protection against both **over-voltage** and current, and extreme-low voltage and current. More complete protection during out-of-control system failures. |
| No External Components Required | Single unit provides space-saving ability as well as cost reduction from extra components. |
| Standard Reset **Threshold** Voltage | Provides consistent and reliable performance in protecting your system from failure during out-of-control system behaviors. |

## Life Cycle

**Figure 5:** Life Cycle Curve for Power Management Components

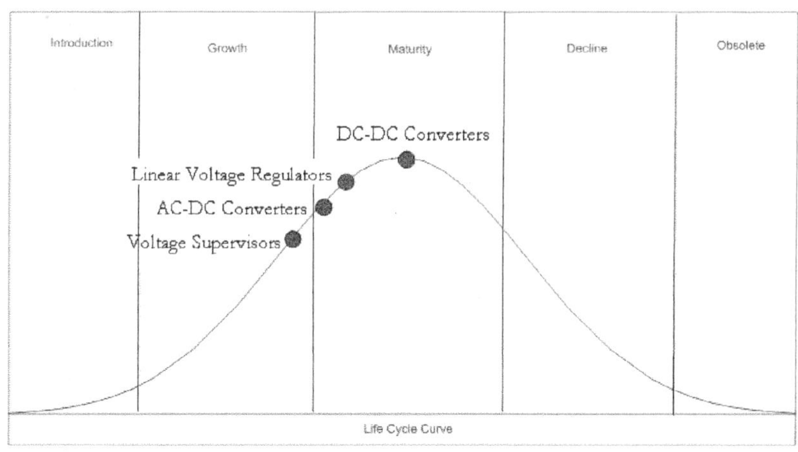


# The Market

## Market Trends

Since power management ICs are a category of semiconductor ICs, the power management IC market faces the same supply and demand shifts. The worst year of semiconductor history was in 2001, with a 19% decline in dollar volume shipments. However, as seen in Figure 6, the market has recovered and is in a good position to grow all the way through 2006[7].

**Figure 6:** Worldwide Shipments for Power Supply and Power Management ICs[7]

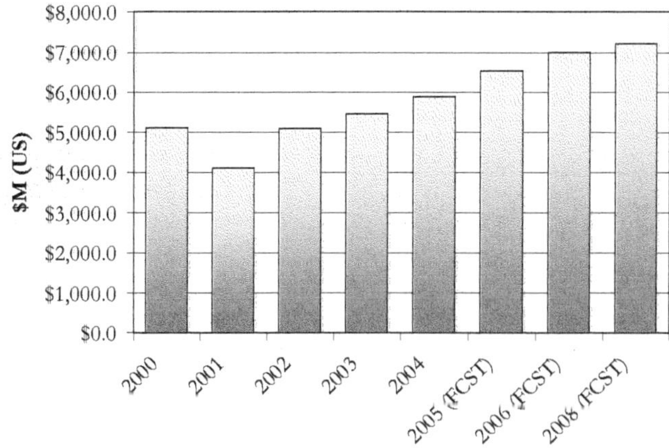

| 2000 | 2001 | 2002 | 2003 | 2004 | 2005 (FCST) | 2006 (FCST) | 2008 (FCST) |
|---|---|---|---|---|---|---|---|
| $5,100 | $4,100 | $5,088 | $5,457 | $5,875 | $6,515 | $6,984 | $7,200 |

**Figure 7:** Worldwide Shipments of AC-DC Power Supplies[6]

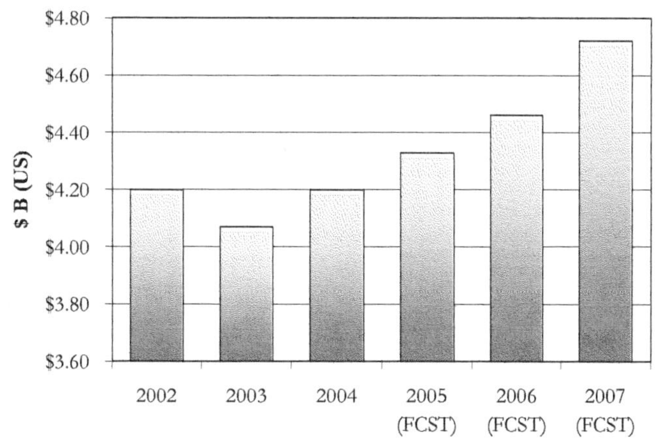


| 2002 | 2003 | 2004 | 2005 (FCST) | 2006 (FCST) | 2007 (FCST) |
|---|---|---|---|---|---|
| $4.20 | $4.07 | $4.20 | $4.33 | $4.46 | $4.72 |

**Figure 8:** Worldwide Shipments of DC-DC Converters[6]

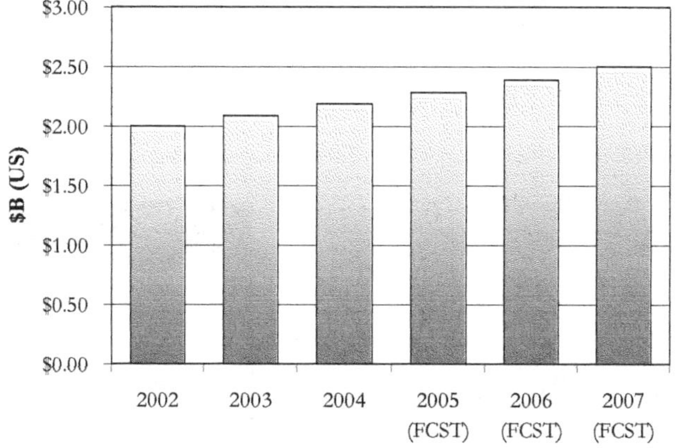


| 2002 | 2003 | 2004 | 2005 (FCST) | 2006 (FCST) | 2007 (FCST) |
|---|---|---|---|---|---|
| $2.00 | $2.09 | $2.19 | $2.28 | $2.39 | $2.50 |

## Regional Segmentation

**Figure 9:** Shipments of Power Supply and Power Management ICs for 2003[7]

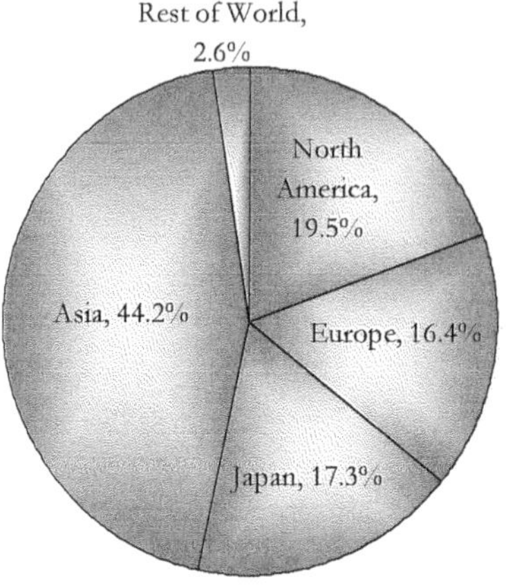


## Market Segmentation

Telecommunications, computers, and peripheral devices are the leading applications with a combined usage percentage of 46.6. According to Venture Development Corporation, the telecom market was "devastated" in 2001 due to the growing concern of the battery life of cellular phones. Computer and peripheral applications have experienced a growth in demand due to the emergence of distributed power architectures and "standby" power supply sections in personal computers and work stations. Notebook computer applications are projected to have the highest growth rate due to the battery-charging and management ICs standards/demands[7].

**Figure 10:** Power Management 2004 Market Segmentation[7]

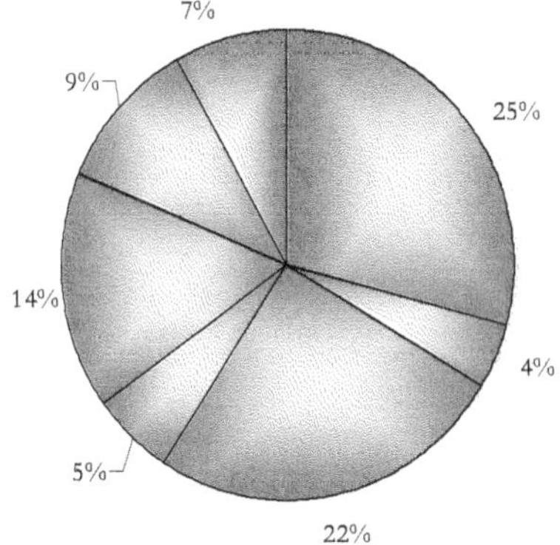


| 2004 Market Segmentation by Application | |
|---|---|
| Telecomm | 25% |
| Computers and Peripheral | 22% |
| Consumer Electronics | 14% |
| Merchant Power Supplies/Chargers | 14% |
| Instruments | 9% |
| Automotive | 7% |
| Datacom | 5% |
| Military/Aerospace | 4% |

# The Players

## Key Players

Since 2002, the top four manufacturers of Power Management ICs have been Texas Instruments, National Semiconductor, Fairchild Semiconductor and Linear Technology. In 2002, the industry was $5,008.1 million with Texas Instruments leading the market as the top supplier. In 2003, Texas Instruments was the top supplier and the market was $5,428.9 million. The lead switched in 2004 when National Semiconductor controlled 12.8% of the market because of a 29.7% growth in revenue in 2003, compared to the average revenue increase of 17.3% for the industry during the same time period. Acquisitions and mergers have played a major role in the shift in vendor shares, specifically the TI/Unitrode and the Fairchild/Samsung acquisitions. In addition to those listed, there are more than 60 vendors with less than 1% market share. Despite the fact that there are approximately 90 participants in this market, the top eight vendors account for more than 50% of the global merchant shipments of these ICs[7].

**Figure 11:** 2004 Key Players in Power Management Market[7]

| | Supplier | Website | Distributors |
|---|---|---|---|
| 1 | National Semiconductor (NSC) | www.national.com | Arrow; Avnet; Digi-Key; Future |
| 2 | Texas Instruments (TI) | www.ti.com | Arrow; Avnet; Digi-Key; Newark InOne; Rochester |
| 3 | Fairchild Semiconductor | www.fairchildsemi.com | Allied; Arrow; Avnet; Digi-Key; Future; Mouser |
| 4 | Linear Technology | www.linear.com | Arrow; Chip Supply; Digi-Key; Linear Express; Minco; NuHorizons; |
| 5 | ON Semiconductor | www.onsemi.com | Avnet; Chip Supply, Minco |
| 6 | Intersil | www.intersil.com | Arrow; Avnet; Chip Supply; Digi-key; Minco; Newark InOne; Rochester |

## References


1 DC to DC Converter. (2005). Retrieved Jul. 27, 2005 from Wikipedia website: <http://en.wikipedia.org/wiki/DC_to_DC_converter>.

2 The Global Markets for AC/DC Switching Power Supplies and DC/DC Converters. (2003). *Venture Development Corporation.* Retrieved Jul. 27, 2005 from VDC Power Conversion and Control White Papers website: <http://www.vdc-corp.com/white/index.html#power>.

3 The Global Market for Power Supply and Power Management Integrated Circuits. (2004) *Venture Development Corporation.* Retrieved Jul. 27, 2005 from VDC Power Conversion and Control White Papers website: <http://www.vdc-corp.com/white/index.html#power>.

4 Integrated Circuit. (2005). Retrieved Jul. 27, 2005 from Wikipedia website: <http://en.wikipedia.com/wiki/Integrated_Circuit>.

5 Linear Regulator. (2005). Retrieved Jul. 27, 2005 from Wikipedia website: <http://en.wikipedia.org/wiki/Linear_regulator>.

6 Microprocessor Supervisor Frequently Asked Questions (FAQs). (2005). Retrieved Jul. 27, 2005 from STMicroelectronics website: <http://www.st.com/stonline/products/families/reset/rst_ans.htm#q1>.

7 Power Management. (2005). Retrieved Jul. 27, 2005 from Wikipedia website: <http://en.wikipedia.com/wiki/Power_management>.

# 4

# Power Supplies

**Power supplies** are electronic devices that are intended to convert an input electrical current into a different output current. Despite what the name may imply, power supplies are not sources of power. They instead take an electrical input and, using modern electronics, convert it into a desired type of current. Power supplies are found in almost all modern electronics mainly due to the use of alternating current (AC).

Alternating current (AC) is the kind of current used nationwide because it is much more efficient to use when distributing power over long distances[2]. Because most electronics would be highly damaged if connected to AC, they usually require direct current (DC). To convert currents from AC to DC, power supplies are often used. While there are many ways to convert electric current, this chapter will focus on three power supply categories: DC power supplies, AC power supplies, and current sources.

**DC power supplies** take an input current and convert it to a desired DC output current. **AC-DC power supplies** accept an AC current, often the standard wall current, and convert it to a specific DC current. These converters are very common and are often integrated into the systems they convert power for such as PCs. Three common voltage outputs of PC AC-DC power supplies are 3.3 volts, 5 volts, and 12 volts. The 3.3-volt and 5-volt ratings are typically used to power the electronic components while the 12-volt supply usually powers things like motors on disk drives and CD-ROMs[1]. DC-DC converters convert one form of DC current into another to meet the requirements of the system.

**AC power supplies** take an input AC signal and regulate it to fit the requirements of the system. AC power sources can usually output variable currents, voltages, and frequencies of AC signal. AC power sources also regulate voltage to eliminate spikes that may occur.
**DC-AC power supplies** convert an input DC signal into an output AC signal, often to convert battery power into AC power, which is frequently used in common household appliances.

**Current sources** take an input signal and convert it to a specific AC or DC signal. Current sources are more "diverse" in that they are capable of producing a wide variety of electrical outputs. They are often used to test and power precision components.

# The Product

## Applications and End Uses

Power supplies are used in any application requiring the modification of electrical current. This makes them necessary for devices such as laptops, motors, and PCs to function. A few uses for power supplies, aside from the common computer sector, include appliances and industrial equipment.

## Product Features and User Benefits

**Figure 1:** Selected Features and their Respective Benefits for Certain Power Supplies

| Product Features | | User Benefits |
|---|---|---|
| Cooled | ⇨ | The device functions better due to cooling, and is at reduced risk of failure due to overheating. |
| Over-Current Protection | ⇨ | Device is capable of preventing large currents from being output despite user inputs, preventing harm to the device, users, and connected components. |
| Over-Voltage Protection | ⇨ | The power supply is capable of limiting output to a preset voltage, which, if exceeded, will shut off the device preventing damage to the supply. |
| Short Circuit Protection | ⇨ | Shorted circuits are limited to a current level in a way similar to over current protection, preventing potential harm to the device. |

## Life Cycle

While still used in testing, troubleshooting, and the like, variable current sources are usually cast aside as an option when compared to the dedicated power supplies (DC/DC, AC-DC and DC-AC in Figure 1). The ability of such devices to be cost-effective doing the job required makes them more appealing.

**Figure 2:** Life Cycle Curve for Selected Power Supplies

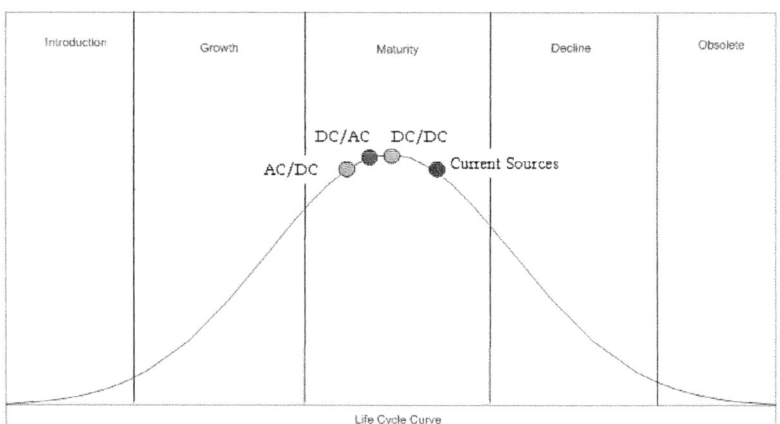


# The Market

## Market Trends

Power supplies are required in almost every situation involving electronic systems. They will likely continue for generations due to the nature of power generation and use in the world today. The nature of AC power requires that power supplies convert the variable-voltage AC power into a steady-voltage DC power to properly run sensitive electronic systems. Over the past few years, global sales of power supplies and Power Management ICs have grown at a steady rate and are forecast to reach nearly $7 billion in 2006.

**Figure 3:** Market for Power Supplies and Power Management ICs[5]

| 2000 | 2001 | 2002 | 2003 | 2004 | 2005 (FCST) | 2006 (FCST) | 2008 (FCST) |
|---|---|---|---|---|---|---|---|
| $5,100.0 | $4,100.0 | $5,088.1 | $5,456.6 | $5,875.4 | $6,515.2 | $6,984.0 | $7,200.0 |

## Regional Segmentation

Asia continues to dominate global shipments of power supplies and Power Management ICs today, with North America, Japan, and Europe far behind in percentage of the total available market. This will likely continue due to the trends of manufacturing moving to the Asia-Pacific region[3].

**Figure 4:** Worldwide Shipments of Power Supply and Power Management ICs[5]

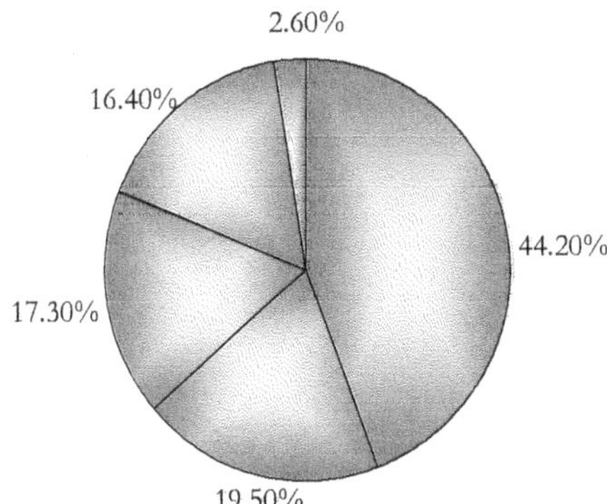


| Regional Segmentation | |
|---|---|
| Asia | 44.20% |
| North America | 19.50% |
| Japan | 17.30% |
| Europe | 16.40% |
| Rest of World | 2.60% |

# The Players

## Key Players

**Figure 5:** Top Ten Power Supply Manufacturers of 2003[4]

| | Supplier | Website | Distributors | Revenue $M (US) |
|---|---|---|---|---|
| 1 | Delta Electronics | www.deltaww.com | | $1,435 |
| 2 | Emerson/Astec Power | www.astecpower.com | All American; Arrow; Digi-Key; Future; Newark InOne, Norvell | $1,030 |
| 3 | Tyco Electronics | power.tycoelectronics.com | Arrow; Avnet; Carlton-Bates; Digi-Key; Future; Mouser; TTI; Sager | $760 |
| 4 | Lite-On Technology | www.liteonit.com | Bell Microproducts; Ingram Micro | $490 |
| 5 | Invensys/Lambda Electronics | www.lambdapower.com | Avnet; Carlton-Bates; Ferrotronic; Jaco; Michigan; Lectrols; Newark InOne; RS Electronics; Simcona; Utech | $440 |

| | Supplier | Website | Distributors | Revenue $M (US) |
|---|---|---|---|---|
| 6 | Acbel Polytech, Inc. (API) | www.acbel.com | Ark Technology | $382 |
| 7 | Marconi Communications | www.marconi.com | GE Access, Tech Data | $375 |
| 8 | Artesyn Technologies | www.artesyn.com | Arrow; Avnet; Digi-Key; Jameco; Mouser; Newark InOne; Norvell; Sager | $314 |
| 9 | Phihong | www.phihong.com | Allied; Mouser; Norvell; | $302 |
| 10 | Shinengen | www.shindengen.com | Coughlin; E-Power Devices; Power Plus; | $282 |

## References


1 Brown, G. (n.d.). How PC Power Supplies Work. Retrieved May 3, 2005 from How Stuff Works website: <http://computer.howstuffworks.com/power-supply.htm>.

2 Direct Current. (2005). Retrieved May 7, 2005 from Wikipedia website: <http://en.wikipedia.org/wiki/Direct_current>.

3 The Global Market for Power Supply and Power Management Integrated Circuits. (2004). Venture Development Corporation.

4 Hammerschmidt, C. (2005). European IC execs debate strengths, weaknesses. Retrieved May 5, 2005 from EE Times website: <http://www.eetuk.com/tech/news/showArticle.jhtml?articleID=60300065>.

5 Mankikar, M. (2005). Top 15 Power Supply Manufacturers in 2003. Retrieved May 2, 2005 from Micro-Tech Consultants website: <http://www.micro-techno.com/top15.html>.

# 5

# Signal Conditioning

Signal conditioners are the "little hidden boxes that manipulate your process signal."[10] There are four different types of signal conditioners:

- Operational Amplifiers (operational amps)
- Comparators
- Timers
- Data Converters

Signal conditioners are embedded in car and truck computers to send warnings when something goes wrong internally. In the design process, "operational amplifiers and comparators are two of the most intricate, and in many ways important, building blocks of an analog circuit."[21]

**Signal conditioning** in simplest terms is the manipulation of electronic data so it can be more easily read by humans or other electronic devices[2]. This function is performed in many different ways by many different devices. Figure 1 shows the major categories in which signal conditioners can be an element.

**Figure 1:** Categories of Signal Conditioners[2]

| Category | Purpose |
|---|---|
| Amplification | Boosts input signal. |
| Isolation | Passes signal from source to measurement device without physical connection. |
| Filtering | Filters out unwanted noise from surroundings. |
| Attenuation | Decreases input signal level. |
| Cold Junction Compensation | Provides accurate thermocouple measurements. |
| Converting | Modifies signal so that it may be read by the system. |

Signal conditioning is most often used in data acquisition applications in which the sensor signals must be converted so they can be interpreted.

**Operational amplifiers (op amps)** are used to boost an A/C signal from its low input level to a higher level that is more easily read by another device[16].

Op amps also keep the integrity of the signal through the amplification. This amplification could be used for the purpose of transmitting the signal or making a signal more readable by a computer. Op amps typically have two input terminals, two supply terminals, one output terminal, and a ground connection[17]. Internally, an op amp uses resistors and a voltage source. Figure 2 denotes the symbol of the typical op amp along with each of it terminals.

**Figure 2:** Operational Amplifier[17]

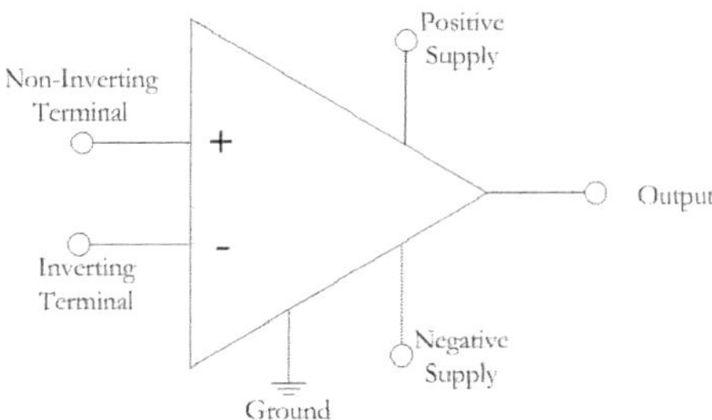


**Comparators** are devices that compare two voltages or currents, and then switch outputs to indicate which of the two voltages or currents is larger[5]. Comparators can also act as a gate by allowing only certain voltages through. If a voltage is higher than the set value, it will pass through without changing; however, if the voltage is below the set value, nothing goes through. The result is that only the transmissions desired go through, serving to block out the unwanted signals. There are many different types of comparators. What typically separates one from another is what the output is desired to be. For example, the LM339 is a dedicated voltage comparator chip that is specifically designed to interface to digital logic. It takes real-world electrical signals and then interfaces them to digital circuitry[5].

The basic difference between op amps and comparators is that op amps are designed to provide accurate and stable output values in closed-loop circuits and comparators are designed to operate in open-loop circuits, meaning that they do not have specified upper or lower voltage limits[13].

**Timers** are a specialized type of clock used to control a sequence of events or a process[22]. Precision timing circuits capable of producing accurate time delays and capable of being configured in two different modes: **monostable multivibrator** (one-shot) or **astable multivibrator** (oscillator). In the time delay or monostable mode of operation, the timed interval is controlled by a single external resistor and capacitor network. In this mode, the timer accepts a signal and is initiated, and cannot change states until the timing cycle completes. In the astable mode of operation, the frequency and duty cycle can be controlled independently with two external resistors and a single external capacitor. Also in the astable mode, the timer has no stable states and therefore changes back and forth (oscillates) between two unstable states without being triggered. The basic purpose is to hold its output line at a high potential for an amount of time after it has been initiated by some signal or pulse.

**Data converters** are devices that change information from one code to another[7]. Data converters are one of the three types of signal conditioners. There are two different kinds of data converters: analog-to-digital converters (A/D Converters) and digital-to-analog (D/A Converters).

An **Analog-to-Digital Converter** is a device that changes continually varying analog signals into a digital signal or binary code that can be used by a computer or other digital devices[4]. Analog-to-Digital Converters can also be abbreviated as ADC, A to D, or simply A/D. The conversion may be done in a single chip or might be just one circuit within a single chip[1]. Usually the analog signal is a voltage and the digital signal is a binary number like a 1 or a 0; however, there are other types of analog signals, including: audio, temperature, energy measurement, synchro/resolver, video decoders, and voltage to frequency converters[1].

ADCs use a process called **Pulse Code Modulation (PCM)** to convert an analog wave signal into a digital signal. PCM samples an analog signal at uniform intervals; every sample is then quantized into a series of binary code[19]. **Differential (Delta) Pulse Code Modulation (DPCM)** is a process that encodes the PCM values as differences between the current and previous values. Compared to PCM, DPCM can dramatically reduce the number of bits required per signal[19]. **Adaptive Differential Pulse Code Modulation (ADPCM)** is a modification of DPCM. It varies the size of the samples and switches the coding scale to compensate for variations in the advanced technique that do not use absolute measurements for each sample point. It also codes the differences frequency and amplitude[18].

Like ADCs, **Digital-to-Analog Converters** are devices that covert a signal, but from digital into analog. D/A Converters can be referred to as DAC, D-to-A, or most commonly D/A. They act as a link between the abstract/digital world, and real life/analog[6]. Many times a simple switch or a network of resisters, current sources, or capacitors may execute this conversion. These are typically used for display or control purposes[9].

# The Product

## Applications and End Uses

The op amp is basically a Universal Analog IC because it can perform most analog tasks. It can function as a line driver, comparator, amplifier, level shifter, oscillator, filter, signal conditioner, actuator driver, current source, and a voltage source[12]. Op amps are used for audio and video pre-amplifiers and buffers as well as voltage clamps. They can function as differential amplifiers, differentiators, integrators, and filters. Precision rectifiers, voltage, and current regulators are also uses. Op amps were originally used for analog calculators to perform basic mathematical operations like addition and subtraction. Op amps are used as oscillators and waveform generators. Finally, op amps can be used as voltage comparators or even as analog-to-digital and digital-to-analog converters[16].

ADCs are used almost anywhere that an analog signal has to be processed, stored, or transported into digital form. ADCs are typically integrated circuits that are required for most digital video editing. They are also used in TV tuner cards and microcontrollers. Digital oscilloscopes require ADCs for applications like software-defined radio. In music, they are used to change analog sound into digital data[4].

DACs are used in some of the same applications as ADCs, such as audio, video, and signal processing. An analog signal from a microphone, instrument, or other sound source can be changed into digital form for storage, editing, transporting, and so on. A video signal from a digital source like a computer must be converted to analog form to be displayed on an analog monitor. DACs are also used in analog signal processing circuits; they have many times replaced potentiometers[9].

The most common applications of timers are to control explosives. Automation processes use timers like home automation and industrial automation. Home automation uses timers for security measures as well as energy conservation. They can be used for lighting, temperature, and audio/video control. Timers are also seen in industrial automation in applications such as controlling machine processes and various safety features[22].

## Product Features and Benefits

**Figure 3:** Product Features and User Benefits of Operational Amplifiers

| Product Features | | User Benefits |
|---|---|---|
| **Rail to Rail** Output | ⇨ | Dynamic range on a low power source. |
| Low Noise | ⇨ | Frequency is consistent. |
| Low Offset Voltage | ⇨ | Device is stable at delivering the voltages it states. Components are matched. |
| Low Input Bias Current | ⇨ | Less power consumed when the device is idle. |

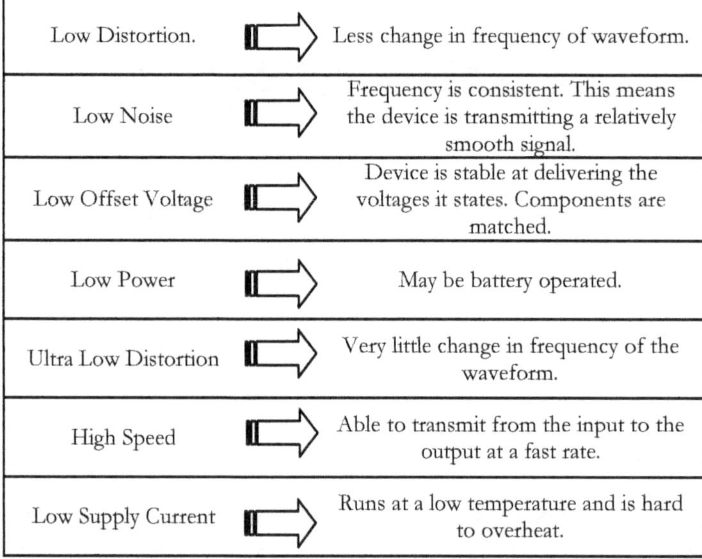

| Low Distortion. | Less change in frequency of waveform. |
|---|---|
| Low Noise | Frequency is consistent. This means the device is transmitting a relatively smooth signal. |
| Low Offset Voltage | Device is stable at delivering the voltages it states. Components are matched. |
| Low Power | May be battery operated. |
| Ultra Low Distortion | Very little change in frequency of the waveform. |
| High Speed | Able to transmit from the input to the output at a fast rate. |
| Low Supply Current | Runs at a low temperature and is hard to overheat. |

**Figure 4:** Product Features and User Benefits of Comparators

| Product Features | User Benefits |
|---|---|
| Propagation Delay | Lower the delay—the time it takes for a change in the input path to change the output path. |
| Low Power Consumption | Requires less voltage and will cost less. |
| Low Input Biasing Current | Less power required to run the device. |
| Low Input Offset Current | Components are matched. |
| Low Input Offset Voltage (max) | Matched components and less variability in voltages. |

**Figure 5:** Product Features and User Benefits of Timers

| Product Features | | User Benefits |
|---|---|---|
| Adjustable Duty Cycle | ⇨ | Allows you to extend the range of workability for the timer. |
| High Stability | ⇨ | High reliability of being free from defects or flaws. |
| Times from Microseconds to Hours | ⇨ | Different timing methods give versatility for timing in different situations. |
| Ease of Use | ⇨ | Can be used by almost everyone without difficulty. |

**Figure 6:** Product Features and User Benefits of Analog-to-Digital Converters

| Product Features | | User Benefits |
|---|---|---|
| Bit Rate | ⇨ | Speed at which signals can be read and identified. |
| Accuracy | ⇨ | More samples taken during modulation create a more accurate simulation of the input. |
| Power Usage | ⇨ | Allows the converter to be designed into a smaller product in which power usage must be a small amount. |
| Number of ADC Inputs | ⇨ | Allow the user to use the converter for more then just one application, keeping the number of converters needed down. |

**Figure 7:** Product Features and User Benefits of Digital-to-Analog Converters

| Product Features | | User Benefits |
|---|---|---|
| Bit Rate | ⇨ | Speed at which signals can be read and identified. |
| Accuracy | ⇨ | More samples taken during modulation create a more accurate simulation of the input. |
| Power Usage | ⇨ | Allows the converter to be designed into a smaller product in which power usage must be a small amount. |
| Number of DAC Outputs | ⇨ | Allow the user to use the converter for more then just one application, keeping the number of converters needed down. |

- **Rail to Rail**
  Rail to rail is common in a CMOS operational amp. Rail to rail provides a dynamic range on a low power source. This means the output voltage can be greatly increased.
- **Noise**
  Noise refers to a random change in the frequency. These devices are also low in noise and therefore frequency is consistent.
- **Offset Voltage**
  Offset voltage is the error that is calculated for unbalanced transistors in the circuits. This is shown when applying a voltage of 0 to a circuit and it ends up with a different output voltage. The input terminals are modeled to account for it. Using the positive input, one is better able to analyze the offset voltage. Offset voltage is low in these, which means that the device does not vary much in transistor consistency and output voltage will not change by much.
- **Input Bias**
  Biasing current is essentially the idle. This is the power needed to run the device. With a lower bias, less power will be consumed. According to Merriam Webster biasing current is the current that establishes a reference for operation.
- **Distortion**
  Distortion is a change in the frequency of the waveform. Therefore these devices have low distortion and are able to maintain a constant frequency.
- **Propagation Delay**
  Propagation delay is the time it takes for a change in the input path to create a change in the output path.

## Life Cycle

**Figure 8:** Life Cycle of Signal Conditioning Products

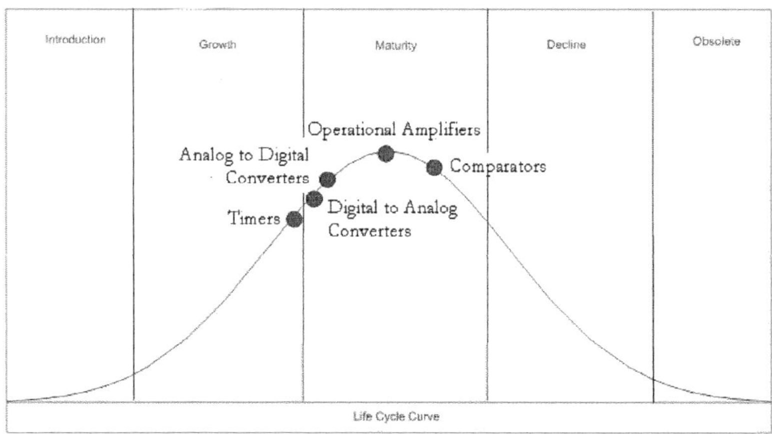

# The Market

## Total Available Market

**Figure 9:** Regional Percentages of Worldwide TAM for Signal Conditioning, $1.5 B (US)[15]

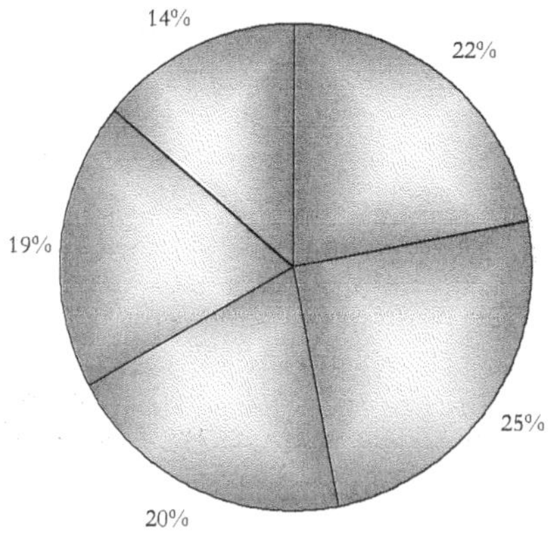


| Regional TAM of Signal Conditioning Products | |
|---|---|
| Asia | 25% |
| North America | 22% |
| Europe | 20% |
| Japan | 19% |
| China | 14% |

The total Analog IC market was $31.4 billion in 2004 and it is estimated to grow to $31.7 billion in 2005. The Amplifier market is about 21% of the total Analog IC Market, therefore in 2004 the amplifier market was about $6.59 billion and is projected to be about $6.66 billion in 2005[11;14]. The operational amplifier figures are based on a 13.4% compound annual growth figure[20].

**Figure 10:** Operational Amplifiers Total Available Market[20]

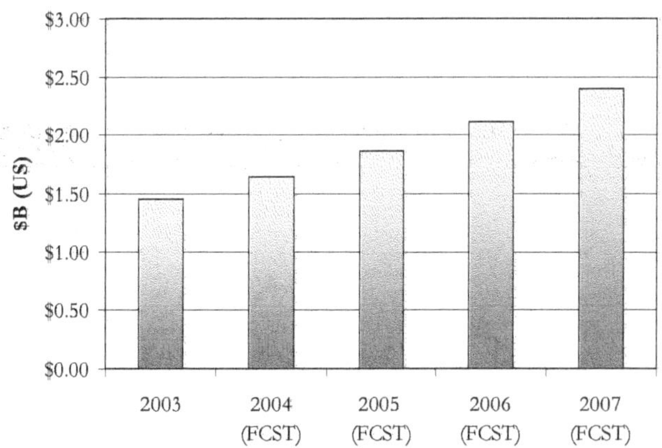


| 2002 | 2003 | 2004 (EST) | 2005 (EST) | 2006 (FCST) | 2007 (FCST) |
|---|---|---|---|---|---|
| $1.28 | $1.45 | $1.65 | $1.87 | $2.12 | $2.40 |

The forecasted trends for comparators are based on 5.3% increase compounded annually[11;14]. By 2009 the total Comparator units shipped is expected to be 2.4 billion.

**Figure 11:** Comparators Total Available Market[11;14]

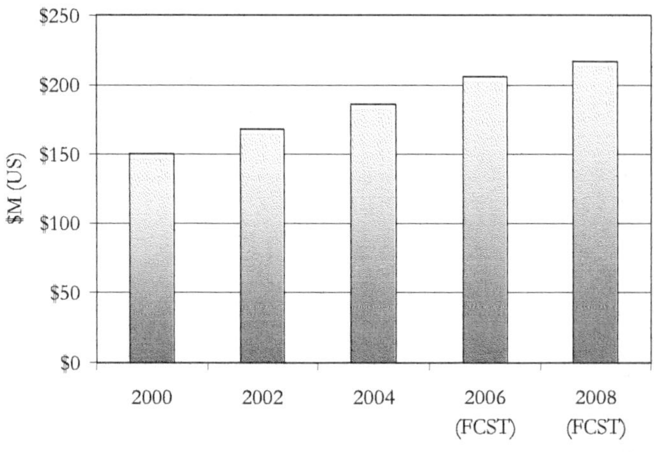


| 2000 | 2002 | 2004 | 2006 (FCST) | 2008 (FCST) |
|---|---|---|---|---|
| $150 | $168 | $186 | $206 | $217 |

**Figure 12:** Total Available Market of Data Converters[11;14]

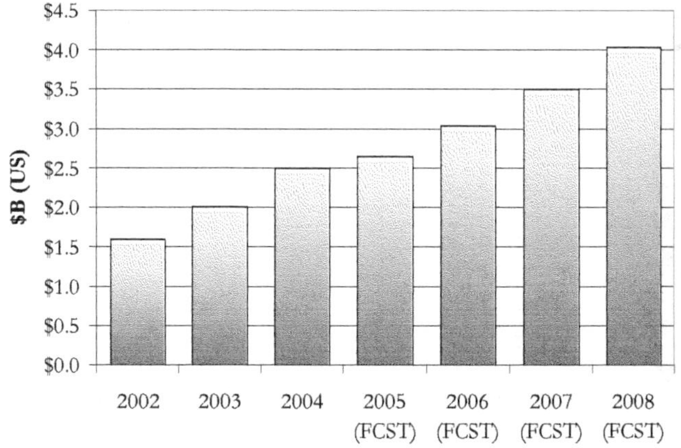


| 2002 | 2003 | 2004 | 2005 (FCST) | 2006 (FCST) | 2007 (FCST) | 2008 (FCST) |
|---|---|---|---|---|---|---|
| $1.6 | $2.0 | $2.5 | $2.6 | $3.0 | $3.5 | $4.0 |

# The Players

## Key Players

**Figure 13:** Top 84% of Manufacturers of Amplifiers and Comparators, $M (US)[15]

| | Supplier | Website | Distributors | Revenue $M (US) | Market Share % |
|---|---|---|---|---|---|
| 1 | Texas Instruments (TI) | www.ti.com | Arrow; Avnet; Digi-Key; Newark InOne | $445 | 17% |
| 2 | Analog Devices | www.analog.com | Arrow; Avnet; Digi-Key; Rochester; Semi Dice | $422 | 16% |
| 3 | National Semiconductor | www.national.com | Arrow; Avnet; Digi-Key; Future; Semi Dice | $413 | 16% |
| 4 | NEC Electronics | www.nec.com | Arrow; Edge; Future Active Components; Future; Jaco; Symmetry | $181 | 7% |
| 5 | Linear Technology | www.linear.com | Arrow; Chip Supply; Digi-Key; Linear Express; Minco; NuHorizons | $143 | 6% |

| | Supplier | Website | Distributors | Revenue $M (US) | Market Share % |
|---|---|---|---|---|---|
| 6 | ST Microelectronics | www.st.com | Arrow; Avnet; Digi-Key; Future; Mouser; NuHorizons | $133 | 5% |
| 7 | NewJRC | www.njrc.com | All American; Digi-Key; Future; Jaco; Mouser; PiK Power | $124 | 5% |
| 8 | Maxim Integrated Products | www.maxim.com | Arrow; Avnet; Digi-Key; Jaco; Dallas Direct (Maxim) | $117 | 5% |
| 9 | Agilent | www.agilent.com | Avnet; Arrow; Digi-key; Future; Newark InOne | $108 | 4% |
| 10 | Renesas Technology | www.renesas.com | All American; Arrow; Avnet; Digi-Key; NuHorizons | $86 | 3% |

**Figure 13:** 2004 Top 10 Manufacturers of Data Converters, Switches, and Multiplexers[15]

| | Supplier | Website | Distributors | Revenue $M (US) | Market Share % |
|---|---|---|---|---|---|
| 1 | Analog Devices | www.analog.com | Arrow; Avnet; Digi-Key; Rochester; Semi Dice | $1,139 | 45% |
| 2 | Maxim Integrated Products | www.maxim.com | Arrow; Avnet; Digi-Key; Jaco; Newark InOne, Dallas Direct (Maxim) | $377 | 15% |
| 3 | Texas Instruments (TI) | www.ti.com | Arrow; Avnet; Digi-Key; Newark InOne | $365 | 14% |
| 4 | Cirrus Logic | www.cirrus.com | Avnet; Newark InOne | $126 | 5% |
| 5 | Linear Technology | www.linear.com | Arrow; Chip Supply; Digi-Key; Linear Express; Minco; NuHorizons | $106 | 4% |

## References


1 About Signal Conditioning. (2005). Retrieved Jul. 11, 2005 from Globalspec website: <http://dta-acquisition.globalsepc.com/LearnMore/Data_Acquisition_Signal_Conditioning/Signal_Conditioning>.

2 A/D Converter. (2005). Retrieved Jul. 12, 2005 from TechEncyclopedia website: <http://www.techweb.com/encyclopedia/defineterm.jhtml?term=A%2FD+converter>.

3 Analog IC growth slows. (2005). Retrieved Jul. 12, 2005 from Reed Business Information website: <http://www.purchasing.com/articles/CA602154>.

4 Analog to Digital Converter. (2005). Retrieved Jul. 12, 2005 from Wikipedia website: <http://en.wikipedia.org/wiki/Analog_to_digital_converter>.

5 Comparator. (2005). Retrieved Jul. 11, 2005 from Wikipedia website: <http://en.wikipedia.org/wiki/Comparator>.

6 D/A Converter. (2005). Retrieved Jul. 12, 2005 from TechEncyclopedia website: <http://www.techweb.com/encyclopedia/defineterm.jhtml?term=D%2Faconverter>.

7 Data Converter. (n.d.) Retrieved Jul. 12, 2005 from Princeton University website: <http://wordnet.princeton.edu/perl/webwn?s=data+converter>.

8 Data Converters from Analog Devices. (n.d.). Retrieved Jul. 12, 2005 from Analog Devices, Inc. website: <http://www.analog.com/analog_root/static/technology/dataconverters/data converters.html>.

9 Digital to Analog Converter. (2005). Retrieved Jul. 12, 2005 from Wikipedia website: <http://en.wikipedia.org/wiki/Digital_to_analog_converter>.

10 Holland, A. (2001). The Little Hidden Boxes that Manipulate Your Process Signals. *Signal Conditioners.* Retrieved Feb. 28, 2005 from Eurotherm website: <http://www.eurotherm.com/training/tutorial/instrumentation/holland/sigcond2.htm>.

11 Inouye, S. (2002). Standard Linear Markets—Worldwide. Retrieved Feb. 27, 2005 from DataBeans Market Research.

12 Mancini, R. (2002). Op Amps for Everyone: Design Reference. Retrieved Jul. 11, 2005 from the Texas Instruments website: <http://www.web-ee.com/primers/files/slod006b.pdf>.

13 Moghimi, R. (2003). Amplifiers as Comparators?. Retrieved Jul. 11, 2005 from Analog Devices, Inc. website: <http://www.analog.com/library/analogDialogue/archives/37-04/comparator.html>.

14 Myson R.B. (2004). Standard Linear Markets—Worldwide. Retrieved Feb. 27, 2005 from DataBeans Market Research.

15 Olsson, M. (2005). Analog Market Revenue Increased in 2004 (G00126978). Retrieved Jul. 11, 2005 from Garter database.

16 Op Amps Tutorial. (n.d.). Retrieved Jul. 11, 2005 from Arizona State University website: <http://enpub.fulton.asu.edu/ece201/flash/opamp.html>.

17 Operational Amplifier. (2005). Retrieved Jul. 11, 2005 from Wikipedia website: <http://en.wikipedia.org/wiki/Operational_amplifier>.

18 PCM. (2005). Retrieved Jul. 12, 2005 from TechEncyclopedia website: <http://www.techweb.com/encyclopedia/defineterm.jhtml?term=PCM>.

19 Pulse-code modulation. (2005). Retrieved Jul. 12, 2005 from Wikipedia website: <http://en.wikipedia.org/wiki/Pulse-code_modulation>.

20 Reilly, N. (2003). Operational Amplifier ICs Exposed (Executive Summary) (SEMC-WW-EX-0314). Retrieved Jul. 11, 2005 from Gartner database.

21 Roubik, G. (1999). Introduction to CMOS OP-AMPS & Comparators. New York, John Wiley & Sons, Inc.

22 Timer. (2005). Retrieved Jul. 18, 2005 from Wikipedia website: <http://en.wikipedia.org/wiki/Timer>.

# Communications

# 6 Fiber Optics

**Fiber optics** is the use of glass or plastic fibers to transmit data using light waves[5]. The beginning of the research into the transmission of light began in 1870 by John Tyndall. Studying a stream of water flowing from one container to a lower one and using light from the sun, he discovered that light uses internal reflection to follow a zigzag path along the stream. Ten years after Tyndall's discovery, Alexander Graham Bell invented the **photophone**, which only carried voices for 200 meters. These were the building blocks of modern fiber optics; however, fiber optic technology[8] did not take off until the 1950s. The exponential growth of technology since then has helped make fiber optics what it is today[8].

Fiber optics obtained its name from the small fiber cables woven together into what is called the **core**. The core is used to transmit light and is surrounded by a coating of glass called the **cladding**. Cladding prevents light from leaking out by reflecting the light back and forth within the core. The core has what is called a "**high refractive index**" which permits light to easily travel throughout its interior. The cladding has a "**low refractive index**" which does not allow light to pass through and thus results in the reflection back into the core. With the reflective capability of the cladding, light rays can still travel through turns and bends in the length of the fiber[8].

**Figure 1:** Basic Fiber Optic Cable[8]

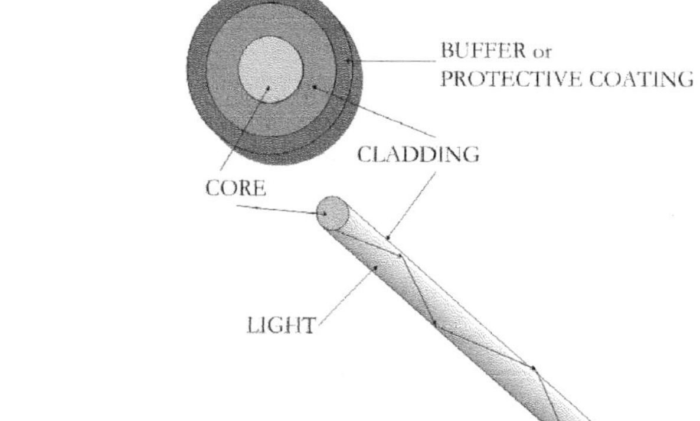

Normally, optic fibers are bundled together within the core, allowing for rays to travel more quickly and in larger amounts. The terminals or product that fiber optic cables are incorporated in depends on the intended application. The three most important parts of an individual fiber within a fiber optic cable are: the **numerical aperture** (ability to reflect light), spectral transmission, and diameter. Light rays that travel through a fiber optic cable are normally in the form of a laser. Lasers are reported to have the capability of carrying 10,000 times the capacity of radio frequencies. However, light has to be used in an encased cladding because it is largely affected by the surrounding environment[8].

Fiber optic products are usually manufactured to carry several standard wavelengths of light. The wavelengths differ depending on quality, distance, and application of the product. For the longest distances of fiber optics cable, a wavelength of 1550 nanometers is used; for shorter wavelengths, 660 nanometers are used. The larger the wavelength usually implies higher performance and distance range that the product can be used, but is also more expensive. Short distance and less complex applications typically use smaller wavelengths due to design simplicity and lower costs[8].

In the fiber optic industry, the combination of bundles, sizes, quality of materials, and wavelength design all have an effect on the overall application. Due to these different characteristics, there is a high complexity for design based on what products are used. Below are four standard fiber optic products that are commonly used in the industry. Though they all have different applications, each product has similar roots of fiber optic operation[8].

A **fiber optic transmitter** is the device that converts electrical signals to optical signals; it includes a source and driving electronics to perform its function[9]. The transmitter utilizes AM, FM, or digital modulation to encode the information with an electrical interface. The transmitter then uses the Light Emitting Diode (LED) or another laser source to input the light signal into the fiber[12].

A **fiber optic receiver** is the device that converts optical signals in to electrical signals; it includes a photodiode semiconductor, signal conditioning circuitry, and an amplifier[9]. The receiver uses one of three types of photodiodes to detect the optical signal, or see the light from the fibers. The three types of photodiodes include[2]:

- Positive-Negative (PN) Junctions
- Positive-Intrinsic-Negative (PIN) Photodiodes
- Avalanche Photodiodes (APD)

Next, the receiver decodes the signal and then sends the electrical signal to a user's device. Therefore, the receiver's job is to decode an incoming signal and convert it to an electronic signal. The transmitter along with the receiver and the optical fiber make up the **optical link**[13].

A **fiber optic attenuator** is a device used to reduce the signal power or amplitude in a fiber optic link by causing a loss without distorting the waveform. When a signal enters the attenuator, the optical fiber absorbs and scatters the signal, emitting a reduced signal. The absorption of the signal is the part of the attenuation process by which the signal is converted to heat. There are several common methods of attenuation including air gaps, micro bending, acousto-optic modulation, and electro-optic modulation. An attenuator's purpose is to step down the strength of the optical signal coming in so the receiver can handle it[1].

**Fiber optic switches** are unique in that they can transmit an optical signal without the usual electronic-to-optical and optical-to-electronic conversions. They are also very versatile because they can interface with both single mode and multimode cable. The main difference between these types of cable is that single mode cable is better for higher bandwidth and longer distances, while the multimode cable allows for the use of an LED light source, which is much less expensive[3].

# The Product

## Applications and End Uses

Fiber optic communication of signals is used extensively in many industries, including the telecommunications, computer, military, and medical industries.

Telecommunications is the number one end user of fiber optics application. Every day, people all over the world access the internet, watch cable television, and use utilities (i.e., lights, electricity) in their homes and offices. Telecommunication uses range from global networks to personal computers. With the growing demand for "bigger, faster, stronger" ways to communicate, whether via the internet, video conferencing, or over the telephone, consumers select the technology that matches their needs. Therefore, the fiber optics industry is responding by inventing more efficient ways to use the internet and new, improved methods to channel surf on TV. All of these advances happen due to the increased use of fiber optic cables, which allow for more information to be processed via the expansion of broadband capabilities[4]

## Product Features and User Benefits

**Figure 2:** Product Features and User Benefits

| Product Features | | User Benefits |
|---|---|---|
| Smaller Diameter; Thinner and Lighter than Copper | ⇨ | Installations are easier and require less space than copper. |
| High Tensile Strength, Durable and Flexible | ⇨ | Good for medical uses; can be maneuvered easily. |
| Security | ⇨ | Less leakage; line tapping is easily detected; good for government bodies, banks, and other top security organizations. Eliminates the need for encryption. |
| Low Hazard Rate, Dielectric | ⇨ | No chance for electrical shock or corrosion. |
| Low Cost | ⇨ | Far less expensive than copper. |
| Longer Service Life | ⇨ | Requires less service from provider and will last longer than copper. |
| Higher Bandwidth in a Smaller Space | ⇨ | Allows for 500 cable channels instead of 100 in far less area used. |
| Less Signal Degradation | ⇨ | Signal loss is less than copper and it is more reliable. |
| Low Power | ⇨ | Require less power to operate, which results in savings by users. |
| Digital Signals | ⇨ | Cables can support computer networks and broadband signals. |

## Life Cycle

The life cycle of fiber optics is determined by the technology. The technology is categorized by Channel Data Rates of 10 mega-bits per second and moves to 1 giga-bit per second and finally to 1 tera-bit per second. Figure 3 describes the three data rates and what each of them equal in terms of bits[8]. The other categorization is **WDM** Channel count, which is **Wavelength-division multiplexing**, or sending several signals through one fiber with different wavelengths of light[8].

**Figure 3:** Channel Data Rates

| 1 mega-bit | = | 1 million bits |
|---|---|---|
| 1 giga-bit | = | 1 billion bits |
| 1 tera-bit | = | 1 trillion bits |

**Figure 4:** Life Cycle Curve for Fiber Optics

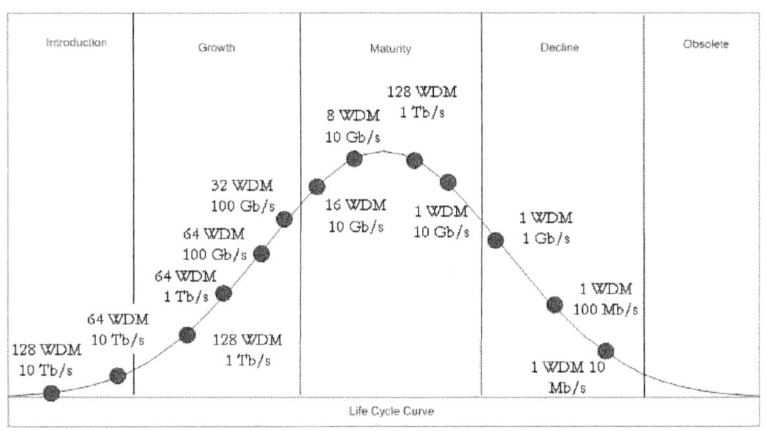


# The Market

## Market Trends

**Figure 5:** U.S. Fiber-To-The-Home Spending $M (US)[14]

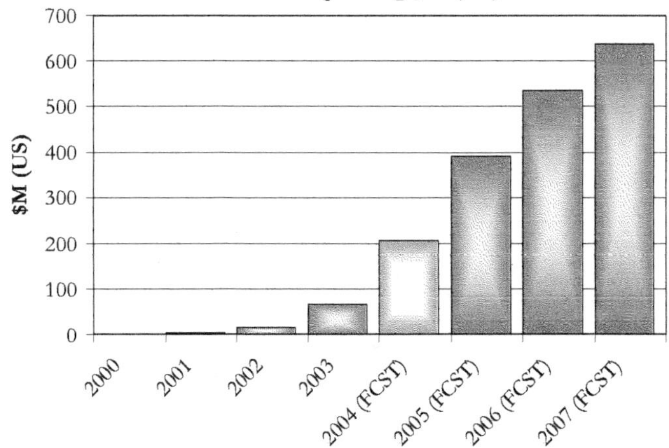

| 2000 | 2001 | 2002 | 2003 | 2004 (FCST) | 2005 (FCST) | 2006 (FCST) | 2007 (FCST) |
|---|---|---|---|---|---|---|---|
| 0 | 3 | 15 | 67 | 206 | 391 | 535 | 636 |

**Figure 6:** North American Optical Fiber Shipments[10]

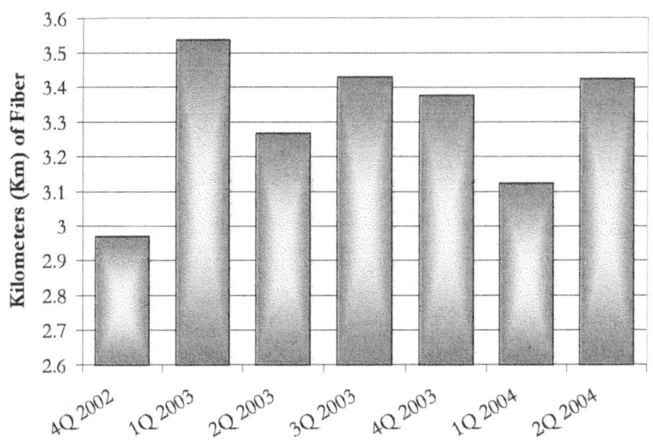


| 4Q 2002 | 1Q 2003 | 2Q 2003 | 3Q 2003 | 4Q 2003 | 1Q 2004 | 2Q 2004 |
|---|---|---|---|---|---|---|
| 2.97 | 3.539 | 3.268 | 3.431 | 3.377 | 3.122 | 3.426 |

**Figure 7:** North American Homes Connected with Fiber-To-The-Home (FTTH)[11]

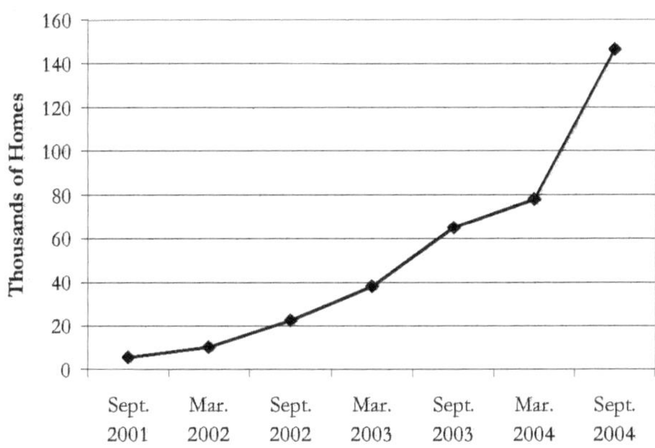

| Sept. 2001 | Mar. 2002 | Sept. 2002 | Mar. 2003 | Sept. 2003 | Mar. 2004 | Sept. 2004 |
|---|---|---|---|---|---|---|
| 5.5 | 10.35 | 22.5 | 38 | 64.7 | 78 | 146.5 |

## Regional TAM Segmentation

**Figure 8:** Worldwide Fiber Demand for 2004, 55 Million Fiber-Kilometers[7]

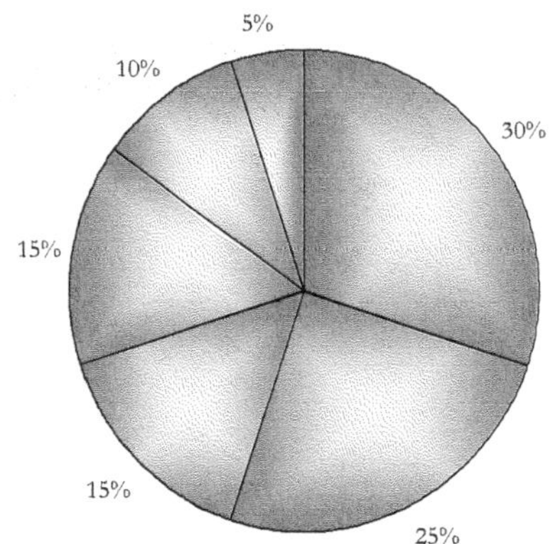


| World Wide Fiber Demand for 2004 | |
|---|---|
| North America | 30% |
| China | 25% |
| Western Europe | 15% |
| Japan | 15% |
| Other Asia | 10% |
| Rest of World | 5% |

## Market Segmentation

**Figure 9:** Worldwide Fiber Demand by Application[6]

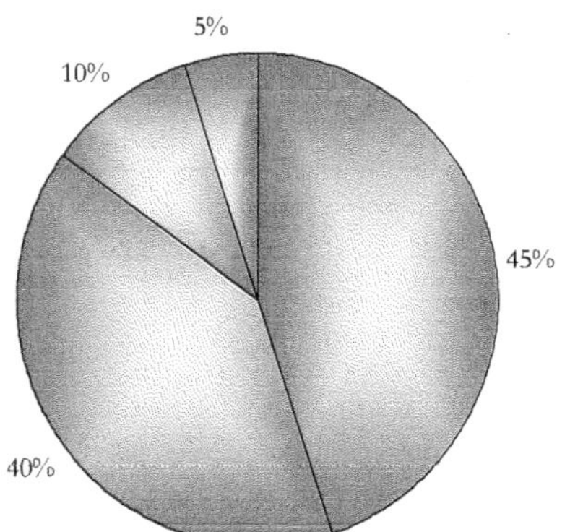


| Worldwide Fiber Demand by Application | |
|---|---|
| Access | 45% |
| Metro | 40% |
| Long Haul, Submarine | 10% |
| Premises | 5% |

# The Players

## Key Players

**Figure 10:** Key Players in Fiber Optics Market

| | Supplier | Website | Distributors | Market Share % |
|---|---|---|---|---|
| 1 | Agilent | www.agilent.com | Arrow; Avnet; Digi-Key; Future; Newark InOne | 26% |
| 2 | Corning | www.corning.com | Accu-Tech, Alltel Communications, Anixter, CSC, Data World; DC Connections; Graybar; Power & Telephone Supply; Sprint North; TVC; WireXpress | 20% |
| 3 | JDS Uniphase | www.jdsu.com | Arrow; Avnet; Bell Microproducts | 16% |
| 4 | Tyco Electronics | www.tyco.com | Arrow; Avnet; Premier Farnell; TTI | 14% |
| Other | | | | 24% |

## References

1 About Fiber Optic Attenuators. (2005). Retrieved Jul. 08, 2005 from Globalspec website: <http://fiber-optics.globalspec.com/LearnMore/Optics_Optical_Components/Fiber_Optics/Fiber_Optic_Attenuators)>.

2 About Fiber Optic Receivers. (2005). Retrieved Jul. 08, 2005 from Globalspec website: <http://fiber-optics.globalspec.com/LearnMore/Optics_Optical_Components/Fiber_Optics/Fiber_Optic_Receivers>.

3 About Fiber Optic Switches. (2005). Retrieved Jul. 08, 2005 from Globalspec website: <http://fiber-optics.globalspec.com/LearnMore/Optics_Optical_Components/Fiber_Optics/Fiber_Optic_Switches>.

4 Applying Fiber Optic Technology. (2005). Retrieved Jul. 08, 2005 from Globalspec website: <http://www.fiber-optics.info/system-designTOC.htm>.

5 The Fiber Guide. (n.d.). Retrieved Jul. 08, 2005 from Communication Specialties, Inc. website: http://www.commspecial.com/fiberguide.htm

6 Fiber Optics. (2002). Retrieved Jul. 07, 2005 from Webopedia website: <http://www.webopedia.com/TERM/f/fiber_optics.html>.

7 Global fiber optics demand by application type in percentages for 2004. (2005). *Fiber Optics Weekly Update, 25.* Retrieved Jul. 08, 2005 from TableBase database.

8 Global fiber optics demand for North America, Western Europe, Japan, China, rest of Asia, and rest of world in percentages and total kilometers for 2004. (2005). *Fiber Optics Weekly Update, 25.* Retrieved Jul. 08, 2005 from TableBase database.

9 Goff, D.R. (2002). A Brief History of Fiber Optic Technology. *Fiber Optic Reference Guide, 3rd ed.* Retrieved from Fiber Optics Info website: <http://www.fiber-optics.info/fiber-history.htm>.

10 Illustrated Fiber Optic Glossary. (2005). Retrieved Jul. 08, 2005 from Fiber Optics Info website: <http://www.fiber-optics.info/glossary.htm>.

11 North American fiber-to-the-home services by number of households in six-month intervals from September 2001 to September 2004. (2005). *Fiber Optics Weekly Update, 25.* Retrieved Jul. 08, 2005 from TableBase database.

12 North American Optical Fiber shipments by fiscal quarter in kilometers for fourth quarter 2002, all quarters 2003 and through second quarter 2004. (2005). *Fiber Optics Weekly Update, 25.* Retrieved Jul. 08, 2005 from TableBase database.

13 Parts of a Fiber Optic Linc. (2005). Retrieved Jul. 08, 2005 from Fiber Optics Info website: <http://www.fiber-optics.info/articles/parts-of-link.htm>.

14 United States annual spending on fiber optics networks for home use in dollars for the period 200 to 2003 and forecast for 2004 to 2007. (2004). *Fiber Optics Weekly Update, 24.* Retrieved Jul. 08, 2005 from TableBase database.

# 7
# Microwave

**Microwaves** are electromagnetic waves that represent a small portion of the wavelength continuum, which can be from 1 millimeter to 30 centimeters. Each microwave has a few certain key characteristics: a specific wavelength, a specific frequency, and a certain amount of energy[25]. Microwave components can be broken down into five key products:

- Oscillators
- Modulators
- Demodulators
- Antennas
- Filters

An **oscillator** is an electrical circuit that produces electrical waves at specified frequencies[11]. The key concept of an oscillator can be related to the pendulum of a clock, which is a type of mechanical oscillator. The pendulum swings at a specific frequency, determined by the number of times the pendulum swings from a point and back to that point in a given amount of time. For a clock, the frequency primarily depends on the length of the pendulum. For any oscillator, energy needs to be converted from one form to another. The pendulum of a clock initially uses potential energy that is converted to kinetic energy, and finally converted back to potential energy[5].

When choosing oscillators, it is important to distinguish which features provide which benefits. It is common for an amplifier to be used to strengthen the signals of oscillators if they are not strong enough for the application[5].

There are several different types of oscillators, including XO (clock), Voltage Controlled Crystal Oscillators (VCXO), Temperature Compensated Crystal Oscillators (TCXO), Oven Controlled Crystal Oscillators (OCXO), and Digitally Temperature Compensated Crystal Oscillators (DTCXO). When deciding between different oscillators, the key performance parameters that must be considered are oscillation frequency, total frequency stability, and connector type[3].

A **modulator** is a circuit or device that combines two signals in such a way that they can be pulled apart later[10]. A modulator in a microwave system converts the output from the oscillator into a form that can be transmitted via antennas. A radio frequency modulator is small piece of electronic equipment that translates the input signal into a RF modulated signal. RF modulators are commonly found in home computers or video game systems that display images on a television. In simple terms, the modulators translate the language of an oscillator into a language that is capable for antennas to understand. In many cases, modulators translate from digital to analog[2].

Once the receiving antenna has obtained the signal from the transmitting antenna, the signals must be put into a form that is capable of producing an output. The **demodulator** serves the reverse purpose of a modulator, which means it typically converts analog signals into digital signals. As with all microwave components, the complexity of a demodulator can range from very simple to containing state-of-the-art technology. Before the signals are processed into the output, it is important that the signals are sent through a filtering system[2].

An **antenna**, when referring to microwave components, is a device that is conductive and designed to radiate RF energy from a transmitter, or to capture RF energy for application to a receiver[6]. For microwave components, antennas serve as the communication unit between modulators and demodulators. Antennas can range from simple to extremely complex, such as a satellite dish.

The most basic type of antenna is the quarter wave vertical. This antenna gets its name because it is a quarter of a wavelength long and is a vertical radiator. It is commonly seen on motor vehicles for two-way communication. Another common type of antenna is the isotropic radiator, which sends and receives signals all around it, similar to that of a light bulb. The isotropic radiator is the antenna of which all other antennas are measured against[3].

An electronic **filter** selectively passes frequency waves from the oscillator into the output. The purpose of the filter is to reduce excess signals that have been contracted along the microwave communication cycle. As with all microwave components, the complexity of filters can range from simple to extremely difficult. When choosing a particular filter, it is important to consider these key parameters: Fc, bandwidth, ripple, insertion loss, and voltage standing wave ratio[1]. Figure 1 shows how the five key components work together in a microwave system.

**Figure 1:** Complete Microwave Communication System[8]

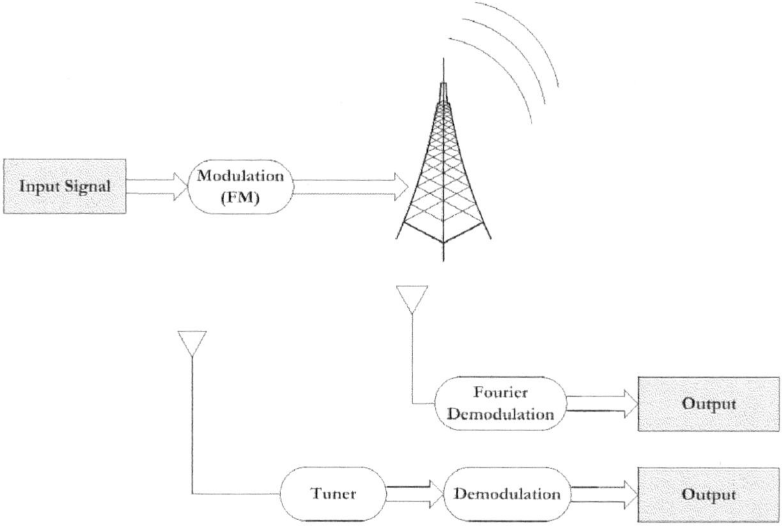


At the beginning of the microwave communication process, the input signal is primarily produced by an oscillator. The oscillator produces a frequency signal that is then received by the modulator. The modulator rearranges the signal into a form that is communicable by the transmitting and receiving antennas. After the receiving antenna receives the signal, it is then transmitted to the demodulator. The demodulator performs the reverse function of the modulator, which means it puts the signal back into a form in which it can be read by an output. Before the signal is sent to the output, it must travel through a filter to expose any improper signals that may have been received by any of the components[16].

# The Product

## Applications and End Uses

Microwave components are primarily used in the telecommunications industry. Modulators and demodulators encode and decode messages sent on waveforms generated by oscillators. Cell phones, walkie-talkies, and routers are examples of devices that use microwave components.

## Product Features and Benefits

**Figure 2**: Features and User Benefits of Modulators and Demodulators

| Product Features | | User Benefits |
|---|---|---|
| Low Noise Floor | ⇨ | No static or noise occurs when transferring frequencies. |
| High Band Frequency Range | ⇨ | Frequency will work with a wide range of devices. |
| Wide Temperature Range | ⇨ | Prevent overheating in high temperatures and freezing in low temperature. |
| Conversion Loss | ⇨ | Higher connectability. |

**Figure 3:** Features and User Benefits of Oscillators

| Product Features | | User Benefits |
|---|---|---|
| Wide Frequency Range | ⇨ | Frequency will work with a wide range of devices. |
| Wide Temperature Range | ⇨ | Prevent overheating in high temperatures and freezing in low temperatures. |
| Frequency Tolerance and Stability | ⇨ | The same intense frequency will remain consistent when working with other components. |
| Shock | ⇨ | The part is durable with products that are used in rugged environments. |

## Life Cycle

It is difficult to determine exactly where microwave components would fit on a product life cycle curve. Microwave components are all dynamic products; manufacturers use older technology to continually develop newer, more efficient, faster, cheaper, and overall better devices and instrumentation.

**Figure 4:** Life Cycle Curve for Microwave Components

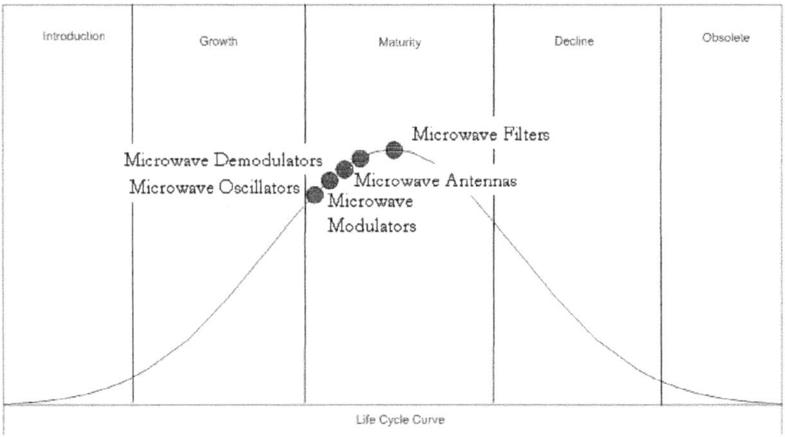

# The Market

## Total Available Market

**Figure 5:** Total Available Market of Microwave Components[18]

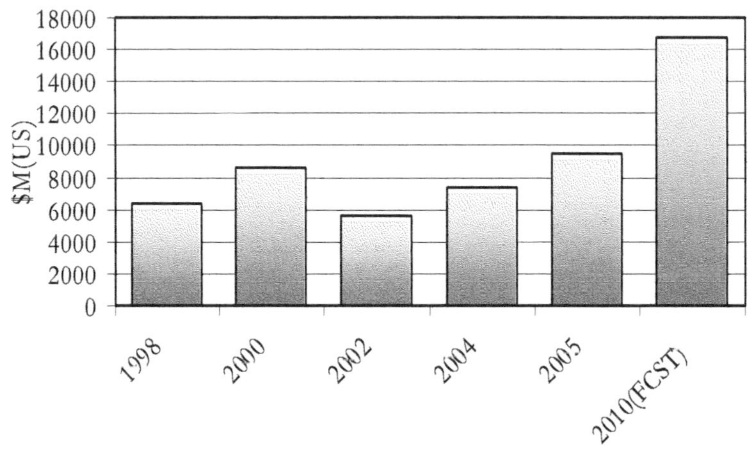


| 1998 | 2000 | 2002 | 2004 | 2005 | 2010 (FCST) |
|---|---|---|---|---|---|
| $6,385 | $8,565 | $5,629 | $7,347 | $9,532 | $16,750 |

As can be seen from figure 5, the market for microwave components has fluctuated throughout the past seven years. From 1998 to 2001, the industry was growing considerably from year to year, due mainly to the escalation of such microwave components in the telecommunications industry as cell phones and radar. In 2001, however, the economy hit a downturn, due mostly to economic uncertainty resulting from 9/11. Along with almost every other industry, the electronics industry suffered heavily. Since 2003, the economy as a whole has picked up and spending on technology has resumed. As a result, the microwave components industry has gained momentum and is projected to have a growth rate of 17.9% per annum until 2010[18].

## Market Segmentation

**Figure 6:** Market Segmentation of Microwave Components[24]

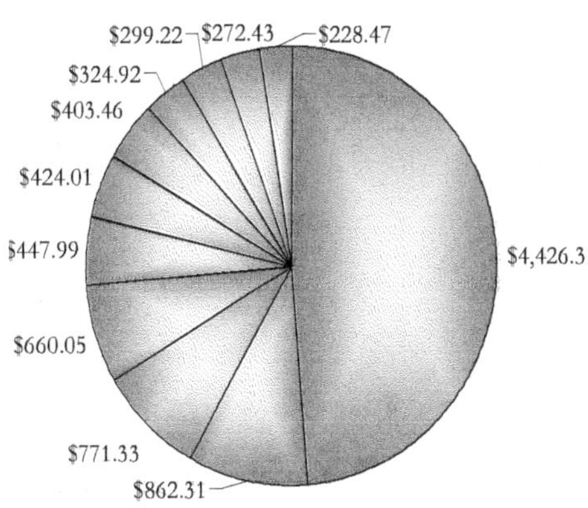


| Market Segmentation 2003 | |
|---|---|
| Cellular/PCS handsets | $4,426.31 |
| Cellular/PCS Base Stations | $862.31 |
| Wave Radio | $771.33 |
| Satellite | $660.05 |
| Consumer and Industrial | $447.99 |
| Cordless | $424.01 |
| WAN Mobile Data | $403.46 |
| Paging | $324.92 |
| GPS Receiver | $299.22 |
| WAN Fixed Data | $272.43 |
| Wireless LAN | $228.47 |
| **TOTAL MARKET** | **$9,120.50** |

# The Players

## Key Players

**Figure 7:** Key Players

| | Supplier | Website | Distributors | Market Share % |
|---|---|---|---|---|
| 1 | Harris | www.microwave.harris.com | AF Comm Supply; Alliance; Hutton; Marketronics; Primus; Talley; Teleiwre; Tessco | 40% |
| 2 | Andrew | www.andrew.com | | 26% |
| 3 | P-Com | www.p-com.com | | 9% |
| 4 | MTI | www.mti.com | | 8% |
| 5 | Remec | www.remec.com | | 6% |

## References


1 6146 500 MHz to 18 GHz pulse modulator. (2005). Retrieved Feb. 20, 2005, from Aeroflex website: <http://www.aeroflex.com/products/gentest/microwave/6146.cfm>.

2 About I/Q Modulators and I/Q Demodulators. (2005). *Global Spec*, 1(1), 5- . Retrieved Feb 27, 2005, from about products database.

3 Antennae. (2000). Retrieved Feb. 20, 2005, from Houghton Mifflin Company website: <http://www.answers.com>.

4 Application Guidelines. (2005). *Global Spec*, 1(1), 5-. Retrieved Feb 27, 2005, from product application database.

5 Basic Antenna Concepts. (2001). Retrieved Jul. 01, 2005, from Wireless Island, Ltd. website: <http://www.wirelessisland.net/basic%20antenna%20concepts.htm>.

6 Brain, M. (1998). Oscillators. Retrieved Feb. 20, 2005, from HowStuffWorks website: <http://www.howstuffworks.com>.

7 Charny, B. (2002). Will Wi-Fi Push Satellite Radio Off the Air. Retrieved Feb. 20, 2005, from <http://news.zdnet.com/2100-1009_22-877669.html>.

8 Complete Microwave Communication System. (n.d.). Retrieved Feb. 01, 2005, from <http://139.134.5.123/tiddler2/c12508/fourier2.html>.

9 The development of catV tuner with GAM demodulator (2000) . Retrieved Feb. 20, 2005, from <http://www.auditory.org/asamtgs/asa94mit/2aEA/2aEA3.html>.

10 Facts and figures. (2005). Andrew Corp. Retrieved Feb. 27, 2005, from <http://www.andrew.com/about/facts_figs.aspx>.

11 Glossary. (n.d.). Retrieved Jun. 30, 2005, from Altinex website: <http://www.altinex.com/Support/glossary_center.htm>.

12 Macsound Music Glossary. (2002). Retrieved Jul, 1, 2005, from Macsound website: <http://www.macsounduk.com/music%20glossary.html#O>.

13 Micro Crystal, (2005). Retrieved Feb. 27, 2005, from Clock Oscillators website: <http://www.mcrystal.com/products.aspx?CatalogID= 1&CategoryID=11&ProductID=9>.

14 Microwave Communication, (2004). Harris Microwave. Retrieved Feb. 27, 2005, from <http://www.microwave.harris.com/>.

15 Oscillators. (n.d.). Ecliptek Corporation. Retrieved Feb. 27, 2005, from <http://www.ecliptek.com/oscillators/E11/>.

16 Products. (2005). P-Com. Retrieved Feb. 27, 2005, from <http://www.p-com.com/products/products.html>.

17 Purdie, I. C. (2003). Retrieved Feb. 20, 2005, from <http://electronics-tutorials.com/>.

18 Rajan, M. (2005). Global Revenue of Compound Semiconductor Products (materials, components and subsystems) to Reach Nearly $30 billion by 2009. Retrieved Jul. 01, 2005, from BCC Research website: <http://www.bccresearch.com/editors/RGB-196R.html>.

19 Raltron Electronics, (2005). Product application. Retrieved Feb. 27, 2005, from <http://www.raltron.com/products/clocks/default.asp>.

20 Rolt, Kenneth . "The Fessenden oscillator: History, Electroacoustic Model, and Performance Estimate ." 18 Feb. 1994. Dept. of Ocean Eng., MIT. 20 Feb. 2005. <http://www.auditory.org/asamtgs/asa94mit/2aEA/2aEA3.html>.

21 Stratex Networks, (2003). Profile. Retrieved Feb. 27, 2005, from <http://www.stratexnet.com/about_us/who_we_are/profile/>.

22 Technical excellence in the same time zone. (2004). Harris Corporation. Retrieved Feb. 27, 2005, from <http://www.harris.com/harris/contact/worldwide-operations.html>.

23 Tech, T. (2005). IQ Assembly *Global Spec*, 1(1), 5-. Retrieved Feb 27, 2005, from IQ Assembly database.

24 Wessles. (1998). Global forecast market size for wireless radio frequency and microwave integration addressable integrated circuits, expressed as demand in US dollars by end-use sector for 1998-2003. Retrieved Jul 1, 2005 from TableBase database.

25 Whitlock, L. (n.d.). Glossary. Retrieved Jul. 1, 2005, from StarChild website: <http://www.starchild.gsfc.nasa.gov/docs/StarChild/glossary_level2/glossary_text.html>.

26 "Worldwide Operations." Harris Corporation. 2004. 24 Feb. 2005 <http://www.harris.com/harris/contact/worldwide-operations.html>.

# 8

# Radio Frequency

**Radio frequency (RF)** is any frequency within the electromagnetic spectrum associated with radio wave propagation. When RF current is supplied to an antenna, an electromagnetic field is created that can then propagate through space. Many wireless technologies are based on RF field propagation[8].

The primary components in the development of RF products are oscillators, transmitters, and receivers. **Oscillators** are devices that are used to generate frequency. They are categorized as either electronic circuit or crystal oscillators. Crystal oscillators are the type used for most RF applications.

**Figure 1:** Common Types of Oscillators

| Category | Common Name |
|---|---|
| Clock | XO |
| Voltage Controlled Crystal Oscillators | VCXO |
| Temperature Compensated Crystal Oscillators | TCXO |
| Oven Controlled Crystal Oscillators | OCXO |
| Digitally Temperature Compensated Crystal Oscillators | DTCXO |

The **clock** or **XO Oscillator** is an uncompensated quartz crystal oscillator. It has no frequency correction system to control variations caused by temperature[6].

The **Voltage Controlled Crystal Oscillator (VCXO)** is a quartz crystal controlled oscillator. An external voltage can be applied to the control pin, which will cause the frequency to change in direct proportion to the voltage applied[6].

**Temperature Compensated Crystal Oscillators (TCXO)** differ from the quartz oscillator's natural frequency which has a temperature characteristic that is corrected by the addition of a temperature compensation (thermistor) network. The thermistor detects ambient temperatures and sends a correction voltage to compensate for the frequency deviation associated with temperature changes[6].

In the **Oven Controlled Crystal Oscillators (OCXO)**, the oscillator uses temperature control circuitry to keep it at a precise temperature; and it is typically made up of precision quartz crystal, an oven block, a temperature sensor, a heating element, oven circuitry, oscillator circuitry, and insulating material. The crystal and its elements are placed inside the oven block to guarantee a stable environment for the most demanding applications[6].

When evaluating an oscillator, factor in the oscillator frequency, the stability of frequency, and the connector type. All of the above types are chosen based on the level of frequency stability needed for the application. Precision applications require a more stable frequency.

Whether it is television, radio broadcasting, or wireless technology, a transmitter is used in many different applications and industries. In communications and information processing, a transmitter is a source that transfers information to a receiver. The **transmitter** is an electronic device that creates varying electric currents, encodes sine waves, and broadcasts radio waves. A transmitter is accompanied by an oscillator to create sine waves, which are the simplest form of varying waves and contain audio and video information. A **modulator** is also employed to encode these sine waves. An antenna is used to broadcast them as radio signals[11].

There are hundreds of types of radio frequency transmitters. Selecting the appropriate device requires an in-depth understanding of modulation methods. On-Off key modulation is the simplest form of modulation; it operates by turning the signal on or off. Amplitude modulation, commonly known as AM, varies the height of the carrier wave to create the required information. Frequency modulation, known as FM, causes the desired frequency to separate from the center frequency. ASK, amplitude shift key, transmits data by changing the amplitude of the signal being transmitted. Frequency shift key, FSK, utilizes two or more output frequencies. The last type of modulation, PSK, phase shift key, is also a digital modulation like FSK, and changes the transmitted signal to accompany the base band data signal[11].

Besides understanding modulation techniques, take note of many other considerations when selecting the correct transmitter. They include supply voltage, supply current, RF connectors, and special features the transmitter includes on board—visual and audible alarms or light-indicating diodes (LED) that indicate operating modes and reception strengths. Frequency range switches and transmitters designed for outdoor use are also available. Transmitters come in many different sizes and shapes for different industries. Transmitters that are used with integrated circuits, ICs, utilize surface mount technology or through hole technology[11].

**Radio Frequency receivers** are electronic devices that separate all incoming signals from each other and convert these signals into a desired format: e.g., audio, video, or data[10].

A receiver works by using an antenna to capture the transmitted signals and a tuner to separate the desired signal from all other signals being received. Detectors, like demodulators, transform the signal back to its original format before transmission. There are multiple ways to decode this information; selecting the correct receiver requires understanding modulation methods. Different types of modulation include AM and FM, On-Off key, and others explained above[10].

Most electronic applications require signal transmission, whether it is close range for local radio stations or long distances for satellite radio or PCS applications. For the many different applications, receivers differ in performance specifications including sensitivity and operating frequency. Sensitivity can be defined as the minimum input signal required to produce a desired output. Operating frequency is the range within which it can receive signals. Digital sampling rate is also an important specification that determines how fast the receiver can draw a sample of the signal being transmitted. Other considerations when selecting a radio frequency receiver include: supply voltage, supply current, and different receiver inputs. Similar to transmitters, some receivers include visual or audible alarms to indicate operating modes and reception strengths[10].

# The Product

## Applications and End Uses

Radio Frequency Technology (RF) is used in a variety of telephone applications, such as cordless telephones and cellular technology, as well is in satellite-based cellular communications. RF is also used in broadband wireless access, wireless local area networks, and wireless personal area networks. Many satellite communications systems, such as DirecTV[1] and Dish Network[2], use Radio Frequency. End uses also include avionics systems, electronic navigation systems, and radar—including continuous wave radar, pulse radar, and automotive radar. Electronic warfare and countermeasures use RF technology to listen, search space and time, intercept, and locate radiated energy, and as a countermeasure to jam communications, to deceive electronically, to act as false targets, to disrupt transmissions, and to change tactics[3].

Mechanical RF oscillators include pendulums, stringed instruments, and tuning forks. Electromechanical RF oscillators are used in microphones and loudspeakers[6].

RF transmitters are used with integrated circuits to incorporate surface mount technology. They also are used in basic radios[11].

---

[1] DirecTV is owned by the DirecTV Group, a subsidiary of News Corporation's Fox Entertainment Group, (2003).

[2] Dish Network is owned by Echostar Communications Corporation, (1996).

RF receivers are usually found in Integrated Circuits (ICs). They integrate Surface Mount Technology (SMT) through Hole Technology (THT) and flat pack. For the telecommunications industry, RF receivers are designed to fit in a metal rack that can be easily installed in a cabinet. They are also found in radios and Electronic Article Surveillance systems (EAS), which can be found in retail stores to monitor shoplifting. Finally, RF receivers are used as an alternative to barcodes in inventory management systems[10].

## Product Features and Benefits

**Figure 2:** Product Features and User Benefits for RF Oscillators

| Product Features | | User Benefits |
|---|---|---|
| Wide Variety of Component Types | ⇨ | Usable in many applications. |
| Low Power Consumption | ⇨ | Allows for extended use. |
| Great Frequency Strength | ⇨ | High performance and reliability. |
| Low Cost for Basic Components | ⇨ | Cost savings. |
| Programmable and Customized Frequency Ranges | ⇨ | Tailored signal strength. |
| Multitude of Suppliers | ⇨ | Customer service, competitive pricing, and inventory availability. |

**Figure 3:** Product Features and User Benefits for RF Transmitters[12]

| Product Features | | User Benefits |
|---|---|---|
| Frequency Range | ⇨ | The device can be selected to fit the user's application. |
| Available for High Power and Low Power applications | ⇨ | The ability to use a transmitter with low power consumption for a low power application can reduce the cost of the entire system. |
| Highly Linearity | ⇨ | When measured, the ability to predict results is very high; they follow a straight line. |

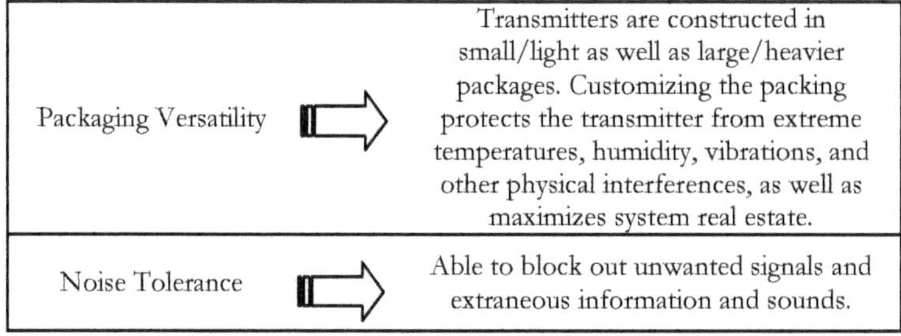

| | |
|---|---|
| Packaging Versatility | Transmitters are constructed in small/light as well as large/heavier packages. Customizing the packing protects the transmitter from extreme temperatures, humidity, vibrations, and other physical interferences, as well as maximizes system real estate. |
| Noise Tolerance | Able to block out unwanted signals and extraneous information and sounds. |

**Figure 4:** Product Features and User Benefits for RF Receivers[12]

| Product Features | User Benefits |
|---|---|
| High Sensitivity | Desired output signal is received with a minimal input signal[10]. |
| High Digital Sampling Rate | High rate at which samples can be drawn from a digital signal (kilo samples/second), which enables application precision[10]. |
| Noise Tolerance | Rejects all unwanted noise including other signals. It does not add any noise or interference to the signal. |
| Packaging Versatility | Customizable Receiver packaging allows for weight and size to be determined, as well as protecting from extreme temperatures, humidity, vibrations, and other physical interferences. In addition, it economizes system real estate. |

## Product Trends

In the industrial segment, the most common trend is Radio Frequency Identification (RFID). **RFID** is a means for storing and retrieving data from remote locations using a small tag or transponder[9]. It is primarily used for product movement in inventory and logistics applications. RFID tags are placed on products that can then be followed through the entire supply chain, while specific information can be updated and changed. RFID tags are also being used in hospitals to prevent infant kidnappings and to scan, store, and modify hospital records to prevent errors[2].

## Life Cycle

**Figure 5:** Life Cycle Curve for RF Components

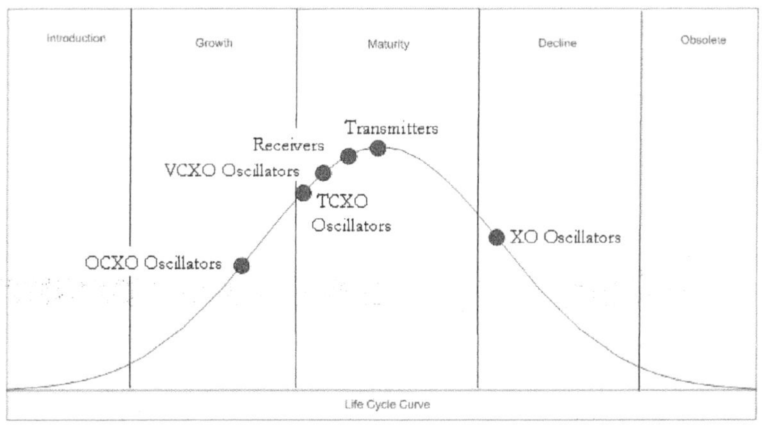


# The Market

## Market Trends

**Figure 6:** RFID Growth and Forecast[13]

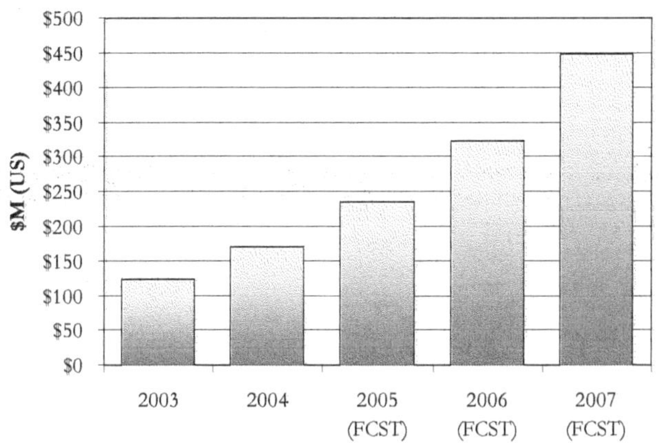


| 2003 | 2004 | 2005 (EST) | 2006 (FCST) | 2007 (FCST) |
|---|---|---|---|---|
| $124 | $170 | $235 | $323 | $448 |

## Market Segmentation

**Figure 7:** Market Segmentation of RFID[5]

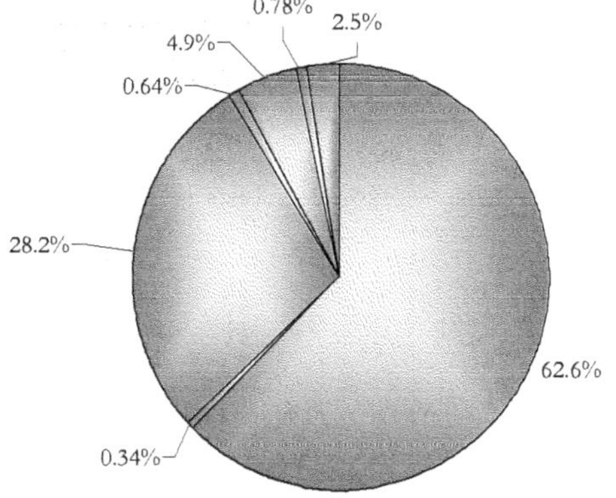


| Market Segmentation for RFID Products for 2004 | |
|---|---|
| Security/Access Control/Purchasing | 62.6% |
| Animals | 28.2% |
| Cartons/Supply Chain | 4.9% |
| Consumer Products | 0.78% |
| Large Freight | 0.64% |
| Humans | 0.34% |
| Other | 2.54% |

**Figure 8:** Market Segmentation of Radio Frequency Semiconductors[1]

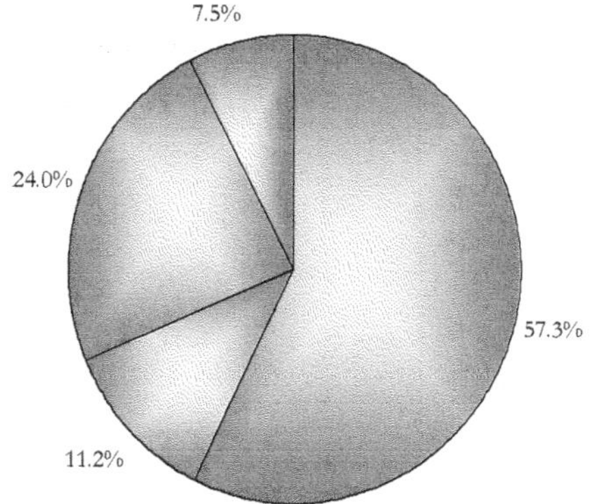


| Market Segmentation of RF Semiconductors | |
|---|---|
| Mobile Communications Equipment | 57.3% |
| Wireless Broadcast Communications | 24.0% |
| Local Wireless Communications | 11.2% |
| Other Wireless Communications | 7.5% |

# The Players

## Key Players

**Figure 9:** Key Players in Radio Frequency Market

| | Supplier | Website | Distributors |
|---|---|---|---|
| 1 | Freescale (Motorola) | www.freescale.com | Arrow; Avnet; Digi-Key; Future; Newark InOne; Richardson; Rochester |
| 2 | Infineon | www.infineon.com | Arrow; Avnet; Digi-Key; Future |
| 3 | Philips | www.semiconductors.philips.com | Arrow; Avnet; Digi-Key; Future |
| 4 | ST Microelectronics | www.st.com | Arrow; Avnet; Digi-Key; Future; Mouser; NuHorizons |
| 5 | NEC | www.nec.com | Arrow; Edge; Future Active Components ;Future; Jaco; Symmetry |

## References


1 Bruederle, S. (1999). RF Semiconductor Market to Reach $8.9 Billion by 2002. Retrieved Jul. 07, 2005 from Gartner database.

2 Deare, Steven. (n.d.). Hospitals go ga-ga over RFID. *Computer World.* Retrieved Feb. 8, 2005 from PC World website: <http://www.pcworld.idg.com.au/index.php/id;484455565;fp;2;fpid;1>.

3 Flikkema, P. (2001). Microwave and RF Product Applications. *The RF and Microwave Handbook*, Chapter.2; pp. 1-61. Retrieved Jul 07, 2005, from ENGnetBASE database.

4 HDTVL. (2005). Retrieved Feb. 20, 2005 from Crutchfield website: <http://www.crutchfieldadvisorcom/ISEO-rgbtcspd/learningcenter/home/tv_hdtv.html>.

5 McGann, R. (2005). RFID Tag Market to Swell Tenfold by 2009. Retrieved from ClickZ website: <http://www.clickz.com/stats/sectors/hardware/article.php/3460851>.

6 Oscillator FAQs. (2004). Retrieved Jul 06, 2005 from International Crystal Manufacturing Company, Inc website: <http://www.icmfg.com/oscillatorfaqs.html>.

7 Radio Frequency. (2005). Retrieved from Wikipedia website: <http://en.wikipedia.org/wiki/Radio_frequency>.

8 RF. (2003). Retrieved Jul. 06, 2005 from Webopedia website: <http://www.webopedia.com/TERM/R/RF.html>.

9 RFID. (2005). Retrieved from Wikipedia website: <http://en.wikipedia.org/wiki/Rfid>.

10 RF Receivers. (2005). Retrieved Jul. 06, 2005 from Globalspec website: <http://rf-components.globalspec.com/LearnMore/Telecommunications_Networking/RF_Microwave_Wireless_Components/RF_Receivers>.

11 RF Transmitters. (2005). Retrieved Jul. 06, 2005 from Globalspec website: <http://rf-components.globalspec.com/LearnMore/Telecommunications_Networking/RF_Microwave_Wireless_Components/RF_Transmitters>.

12 Seely, W. (2001). Circuits. *The RF and Microwave Handbook*, Chapter.2; pp. 1-61. Retrieved Jul 07, 2005, from ENGnetBASE database.

13 Weber, A. (2003). Consumer Products: product development and time to market are more important than ever. *Assembly Mag*, (1050-8171), pg.46. Retrieved Jul 26, 2005, from TableBase database.

# Wireless Networking
## Bluetooth, WLAN, & WWAN

Understanding physical networks is the first part of understanding wireless technology. **Networks** are groups of computers that communicate through wires or through temporary means like a modem[1]. Typical computer networks use coaxial, Ethernet (cat-5), or fiber optic cable to transmit data[3]. A wireless network is basically a network type communication system that does not use cables or wires[5]. Wireless communication and networks use other mediums, such as infrared light, radio waves, and electromagnetic waves to transmit data packets[2]. Figure 1 shows the four basic types of wireless networks.

**Figure 1:** Four Types of Networks[4]

| Network | Acronym | Definition | Range |
|---|---|---|---|
| Wireless Local Area Network | WLAN (Wi-Fi) | Allows users in a local area to gain network access. Local areas include university campuses, libraries, and small businesses. | 0-350 feet |
| Wireless Personal Area Network | WPAN | Primarily made up of Bluetooth technology. Allows users to connect personal devices within 30 feet. Personal devices include laptops, PDAs, cell phones, printers, and scanners. | 0-30 feet |
| Wireless Metropolitan Area Network | WMAN | Allows for multiple networks in a metropolitan area to connect, including between buildings in a city. | 0-165,000 feet |
| Wireless Wide Area Network | WWAN | Network spans over large areas, such as cities or countries. They use satellite systems and antenna sites, also referred to as 2G systems. | 0-35,000 feet |

The following chapters will discuss WLANs and WWANs in addition to the newest WPAN, Bluetooth technology.

## References

1 Computer Network. (2005). Retrieved Jun. 29, 2005, from Wikipedia website: <http://en.wikipedia.org/wiki/Computer_network>.

2 Introduction to Wireless Networking. (2005). Retrieved Jun. 29, 2005, from Webopedia website: <http://www.tutorial-reports.com/wireless/introduction.php>.

3 Local area networks. (2003). Retrieved Jun. 29, 2005, from Webopedia website: <http://www.webopedia.com/TERM/L/local_area_network_LAN.html>.

4 Tabona, A. Z. (2004, May 20). An Introduction to Wireless Networking (Part 1). Retrieved Jun 29, 2005, from Windows Networking website: <http://www.windowsnetworking.com/pages/article_p.asp?id=310>.

5 Wireless. (2005). Retrieved Jun. 29, 2005, from Wikipedia website: <http://en.wikipedia.org/wiki/Wireless>.

# 9

# Bluetooth

**Bluetooth** is a wireless networking technology that uses radio waves. It is designed to connect and simplify the communications between devices within a short range[1]. A form of wireless LAN, it is compatible across a range of very different operating systems and devices. It has a short range, typically less than 30 feet, which acts as a security feature. To intercept data, the interceptor must be physically close to the technology[1]. The founders of Bluetooth include Ericsson, IBM, Intel, Nokia, and Toshiba. A Bluetooth Special Interest Group has been formed to qualify and pass interoperability testing to products with Bluetooth technology.

## The Product

### Applications and End Uses

Bluetooth technology is not only for computer devices; it can be found in headsets, MP3 players, phones, and even PDAs. A Bluetooth-enabled headset allows a call to be answered or sent without actually having a wire to or from the phone. This feature comes with the safety benefit that ensures people will no longer need to place a phone generating 100 times the radio energy close to their head. Not only can the phone communicate with the headset, but it can also communicate with anything that is Bluetooth-enabled. This is helpful in sharing contact information between a cell phone, PDA, and email. With the right software, it could be used to look up a phone number on the computer and call it with a cell phone without ever touching the cell phone. However, the right software must be installed on the cell phone and the computer[2].

Other products that are Bluetooth-enabled include[3]:

Access points (allow Bluetooth-enabled devices to connect to a local area network)

- Fax machines
- Desk top computers
- Laptop
- Printer
- In-car GPS systems
- In-car entertainment systems
- Tablet
- In-car phone systems
- Computer mouse
- CPU
- Computer keyboard

## Product Features and User Benefits

**Figure 1:** Product Features and User Benefits of Bluetooth Technology[4,7]

| Product Features | User Benefits |
|---|---|
| Operates in the **2.56 GHZ ISM Band** | The ISM band is the industrial, scientific, and medical radio band originally reserved for non-commercial use of RF electromagnetic fields. This band is available globally; products that are Bluetooth-enabled can be used beyond the country in which they were purchased. |
| Uses **FHSS** | A transmitter and receiver hop together from one frequency to another in an arranged pattern. Many systems can operate using the same sequence because the chance of collisions is low. However, the chance of collision increases as more users use the same sequence. |
| Can Support up to 8 Devices in a **Piconet** | Allows one master device to interconnect with up to seven slave devices. Including the access point (master), eight devices may be interconnected; (1) access point (2) CPU (3) monitor (4) keyboard (5) mouse (6) printer (7) cell phone (8) PDA. |
| Omni-Directional, Non-Line-of-Sight Transmission Through Walls | Devices connecting using Bluetooth may be pointed at each other in any direction, at any angle to connect. |
| 10 Meters to 100 Meters in Range | Some devices must be within 10 meters (approx. 3 feet) to connect to the desired device, and others must be within 100 meters (approx. 300 feet). |
| Low Cost ($5-$10) | The Bluetooth card can be placed in many different devices without causing the price of those devices to increase dramatically. |
| 1mW Power | Has very low power consumption and therefore devices do not require very large internal batteries. This will hold the size down. |
| Extended Range with External Power Amplifier (100 meters) | Has the ability to extend the specified range to as much as 100 meters. |

## Life Cycle

**Figure 2:** Life Cycle Curve for Bluetooth

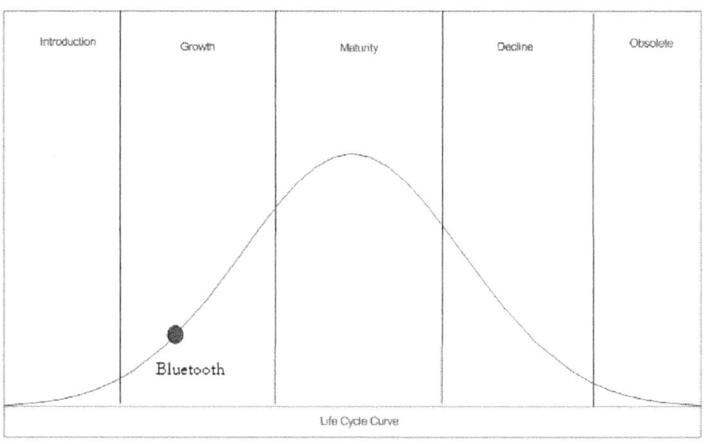


# The Market

## Market Trends

**Figure 3:** Bluetooth Total Available Market[6]

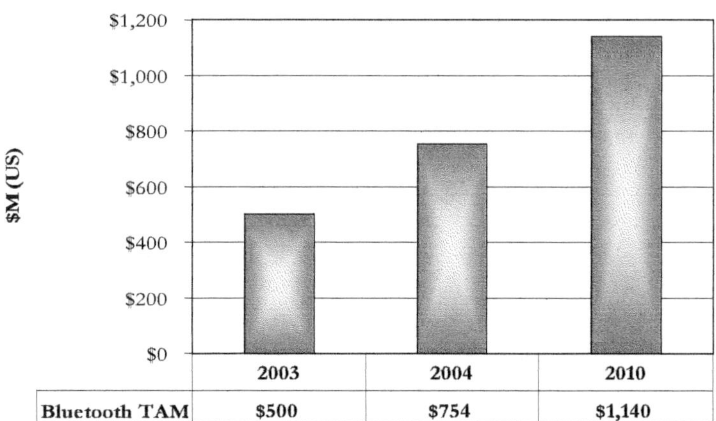


| | 2003 | 2004 | 2010 |
|---|---|---|---|
| **Bluetooth TAM** | **$500** | **$754** | **$1,140** |

In 2003, the majority of sales of Bluetooth units were to buyers in Europe, whereas U. S. buyers only accounted for 10% of the world's Bluetooth units sold. Implementation is higher in Europe due to the recently passed law against cellular phone use while driving. Japan had passed a similar law in

2004[6]. It is expected that Bluetooth-enabled products will experience a 60% CAGR between 2003 and 2008[6].

**Figure 4:** Bluetooth Sales in Billions of U.S. Dollars

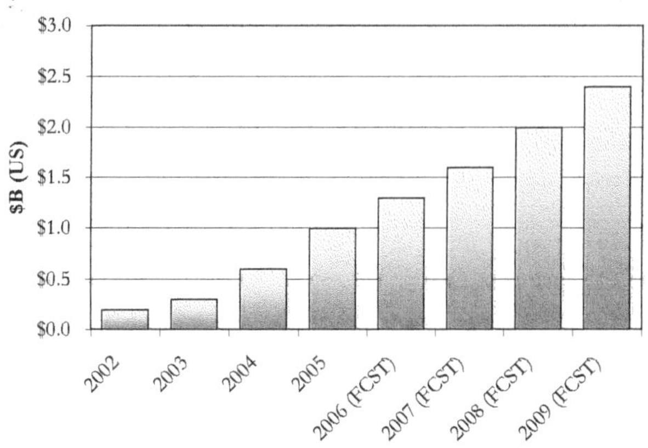


Source: iSuppli

| 2002 | 2003 | 2004 | 2005 | 2006 (FCST) | 2007 (FCST) | 2008 (FCST) | 2009 (FCST) |
|---|---|---|---|---|---|---|---|
| $0.2 | $0.3 | $0.6 | $1.0 | $1.3 | $1.6 | $2.0 | $2.4 |

# The Players

## Key Players

The market for Bluetooth vendors has been consolidating since 2002, with a few key players left. Industry analysts are not revealing market share figures; however, many companies in the industry are estimating their own and competing companies' market shares.

Currently, CSR is the top producer of the Bluetooth chipset. The vice president of CSR Eric Janson has estimated that CSR's market share is 51%. However, Toshiba analysts rank CSR at 40%. Either way, CSR holds the largest share of the market. Broadcom holds a close second to CSR in the North American market. However, it is a toss up for the third, fourth, and fifth positions among Infineon Technologies, Texas Instruments, and RF Micro Devices Inc[6]. Japan-based companies Toshiba and Renesas also hold positions in the market, predictably the sixth and seventh positions.

**Figure 5:** Key Players in Bluetooth Market

| Company | Website | Distributors |
|---|---|---|
| CSR | www.csr.com | Beta Technology Sales; Bonser-Philhower Sales; EIR; J-Squared Technologies; Luscombe Engineering Company; Moultrhop Sales; Ontecssi; Strategic Sales; WestRep. |
| Broadcom | www.broadcom.com | Arrow; Avnet; Bell Microproducts. |
| Infineon | www.infineon.com | ACG North America; Arrow; Avnet; Digi-Key; Future. |
| Texas Instruments | www.ti.com | Arrow; Avnet; Digi-Key; Newark InOne. |
| RF Micro Devices Inc. | www.rfmd.com | |
| Toshiba | www.toshiba.com | Arrow; Avnet; Nu-Horizons; TAEC. |
| Renesas | www.renesas.com | All American; Arrow; Avnet; Digi-Key; NuHorizons. |

## References


1 Bluetooth. (2005). Retrieved Jun. 30, 2005, from Webopedia website: <http://www.webopedia.com/TERM/b/bluetooth.html>.

2 Bluetooth Wireless Networking Explained. (2005). Retrieved Jul. 1, 2005, from the Travel Insider website: <http://www.thetravelinsider.info/roadwarriorcontent/bluetooth.htm>.

3 Franklin, C. (2005). How Bluetooth Works. Retrieved Jun. 30, 2005, from How Stuff Works website: <http://electronics.howstuffworks.com/bluetooth.htm>.

4 Gupta, P. (2005). Bluetooth technology. Retrieved Jun. 30, 2005, from Mobile Info website: <http://www.mobileinfo.com/Bluetooth/competitors.htm>.

5 Lawler, R.. (2004). Bluetooth Application Market Trends. Retrieved Jun. 20, 2005, from FACCTs website: <http://www.faulkner.com/products/faccts/00018685.htm>.

6 Supply Stable, Prices on the Dive. (2005). Retrieved Jul. 01, 2005, from Global Sources website: <http://www.telecom.globalsources.com/gsol/I/Bluetooth-headset/a/9000000063702.htm>.

7 What is Bluetooth? (n.d.). Retrieved Jun. 30, 2005, from Palo Wireless website: <http://www.palowireless.com/infotooth/whatis.asp>.

# 10

# WLAN

## Wireless Local Area Networks

**Wireless Local Area Networks (WLANs or Wi-Fis)** are local area networks that use radio waves instead of wires or cables to communicate between its notes[15]. Many Wireless LAN technologies have fizzled out, and other promising technologies are currently being hyped but are months or years away from launch. Currently, the 802.11 standard is by far the most established and preferred WLAN chipset[1]. The other Wireless LAN products that can be found are Bluetooth, and HomeRF.

**802.11** is a group of specifications for WLAN technology created by the IEEE (Institute of Electrical and Electronics Engineers); it details a connection from a wireless user and a base station or between two wireless users[1]. 802.11 was initially created to provide wireless networks in large companies. However, it is now the most common form of wireless connection for all consumers.

There are many 802.11 chipsets, such as 802.11a (54 Megabit), 802.11b (11 Megabit), 802.11g (54/11 Megabit), and 802.11e, g, h, l, n, which are very similar to the a/b/g chipsets but with added spectrum management, or Quality of Service and VoIP enhancements. For the average consumer, the a, b, and g chipsets are the staples[1].

**Wi-Fi** or Wireless Fidelity is a term that is used when referring to any type of 802.11 networks. Some 802.11 products are tested, approved, and certified by the "Wi-Fi Alliance," which means that the certified product can use any brand of access point with any other brand of client hardware. However, many times a Wi-Fi product will work with any other Wi-Fi product, whether or not each is certified[16].

The **802.11a** operates at 54 Megabits and it is considered to be "Wi-Fi Certified." This chipset, as compared to the b or g, has a better chance of withstanding RF interference than the 802.11b. It also provides better support for multimedia, voice, video, and large-image applications in populated user situations. The downside is that the 802.11a is not compatible with the 802.11b and has a shorter range[1].

The **802.11b** is also considered to be "Wi-Fi Certified" and it operates at 11 Megabits. This chipset does not require as many access points and it offers "high-speed" access to information from up to 300 feet from the base station[1].

The **802.11g** has replaced both the 802.11a and the 802.11b. It operates at 54 Megabits, like the 802.11a; however, it is compatible with the 802.11a and "backwards compatible" with the 802.11b. The 802.11g has better security features and commonly replaces the 802.11b[1].

**HomeRF**, Home Radio Frequency, was created to provide wireless networks in home, not businesses. It was designed as a cost-effective way for home users to participate in wireless technologies. Its range is only 150 feet from the base station. No vendors or working groups are currently supporting HomeRF anymore[9].

# The Product

## Applications and End Uses

The primary use of wireless LANs is for internet access and networking for a company, a school, a home, or even the public that extends beyond a company's firewall[5]. WLANs are now found at "**wireless hotspots**," which include airports, libraries, hotels, and even restaurants[10]. A **wireless hotspot** is a place for users to have an access point to a wireless network. WLAN chips are found in laptop computers, desktop computers, PDAs, Softphone Client software, handsets, cellular handsets, and even custom devices, such as scanners[14].

## Product Features and User Benefits

**Figure 1:** Wireless LAN Features and User Benefits

| Product Features | | User Benefits |
|---|---|---|
| Coverage | ⇨ | Can use laptops and other wireless-enabled devices in many office buildings, campus buildings, and even public hotspots. Makes real-time data available in more areas. |
| Ethernet Network | ⇨ | All computers are enabled with Ethernet connections and many with Ethernet **NIC cards** as well. (Network Interface Connection card, required to make connection to Wireless Network) |
| High Speed | ⇨ | Speeds are as fast or faster than a cable hookup. |

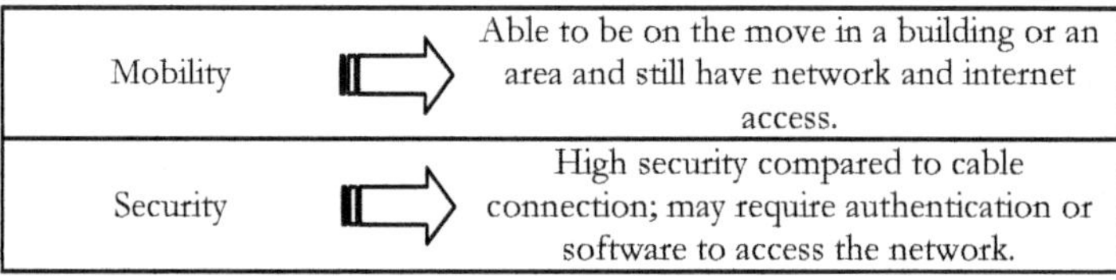

| | | |
|---|---|---|
| Mobility | ⇨ | Able to be on the move in a building or an area and still have network and internet access. |
| Security | ⇨ | High security compared to cable connection; may require authentication or software to access the network. |

## Product Trends

The trend in 802.11 technologies is leaning away from the 802.11a. Few consumers have implemented the 802.11a or have plans to implement it in the future[14]. In 2006, 802.11n technologies will be released. However, before that consumers will be able to purchase products that support 802.11g, 802.11a, and 802.11n technologies[2]. The 802.11i technology has also been standardized and will be seen in future endproducts[2].

## Life Cycle

Both the 802.11b and the 802.11a were introduced in late 2001. The 802.11b became the most popular and best-selling chipset. It is still in high demand. In mid-2003, the 802.11g was introduced to the market. It is backwards compatible with 802.11b and is still increasing in market acceptance. In 2006, the 802.11n will be introduced to the market. It will be four to five times faster than the 802.11g and could be up to 40 to 50 times faster than the 802.11b[1].

**Figure 2:** Life Cycle Curve for WLAN[11]

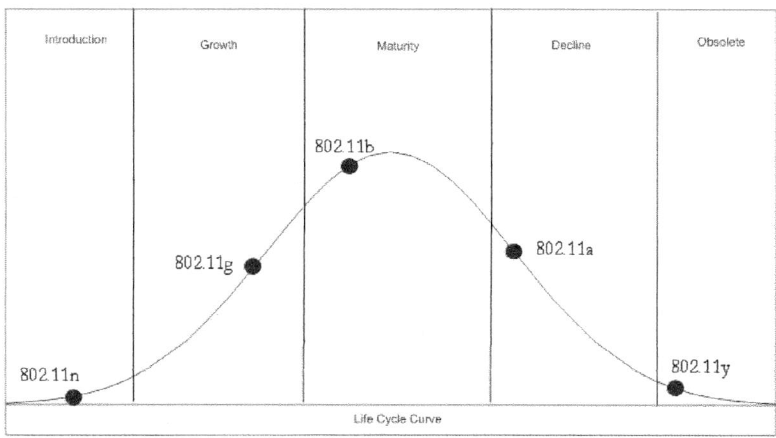

# The Market

## Total Available Market

**Figure 3:** WLAN Equipment Revenues[17]

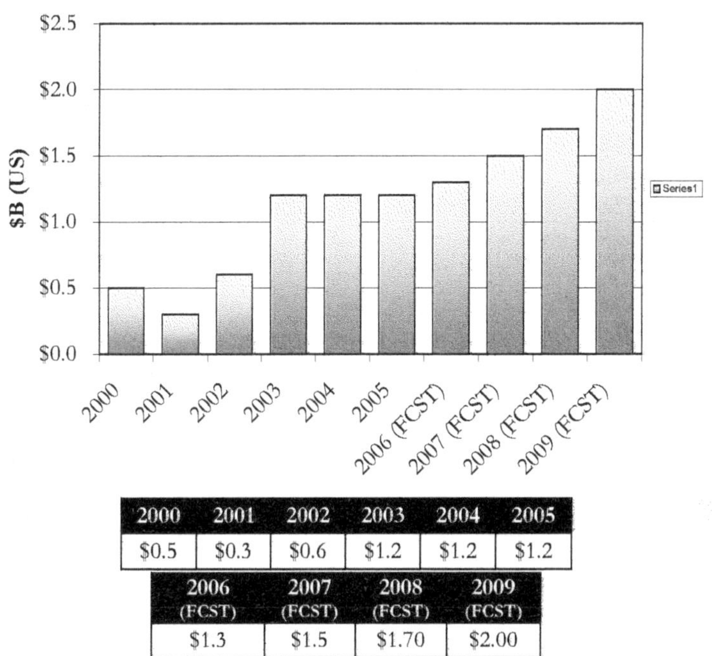


| 2000 | 2001 | 2002 | 2003 | 2004 | 2005 |
|---|---|---|---|---|---|
| $0.5 | $0.3 | $0.6 | $1.2 | $1.2 | $1.2 |

| 2006 (FCST) | 2007 (FCST) | 2008 (FCST) | 2009 (FCST) |
|---|---|---|---|
| $1.3 | $1.5 | $1.70 | $2.00 |

Source: iSuppli Corporation

**Figure 4:** Segmentation of TAM by Individual Product[17]

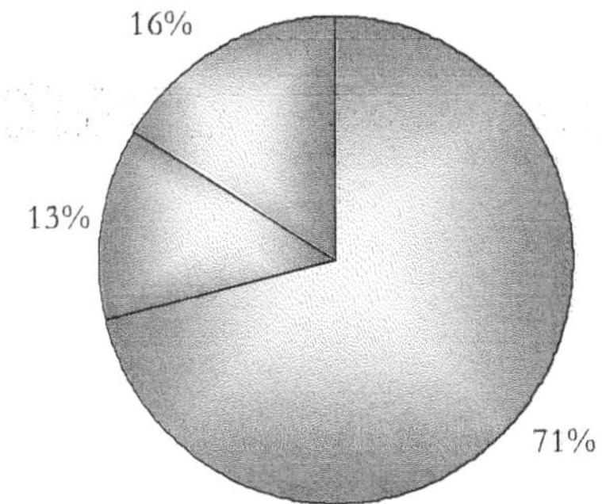


| Segmentation of TAM by Individual Product | |
|---|---|
| Access Points | 71% |
| NICs | 13% |
| Infrastructure (switches, appliances, controllers, mesh networking gear) | 16% |

## Market Segmentation

The market segmentation has been divided into three common user groups: mobile PCs, handheld PCs, and 2.5, 3G phones. (The mobile PC category includes laptops, and the handheld PC category includes PDAs. 2.5, 3G phones are third generation cell phones and other handsets.)

**Figure 5:** Forecasts for Mobile Products with WLAN[8]

| Thousands of Units | 2001 | 2002 | 2003 | 2004(FCST) | 2005(FCST) | 2006(FCST) |
|---|---|---|---|---|---|---|
| Mobile PCs | 803 | 3,254 | 6,685 | 9,985 | 13,192 | 16,542 |
| Handheld PCs | 94 | 488 | 1,685 | 4,126 | 8,846 | 0 |
| 2.5, 3G Phones | 0 | 0 | 3,388 | 9,525 | 22,053 | 42,706 |

## Regional Segmentation

**Figure 6:** Segmentation of WLAN Revenues by Geographic Region[17]

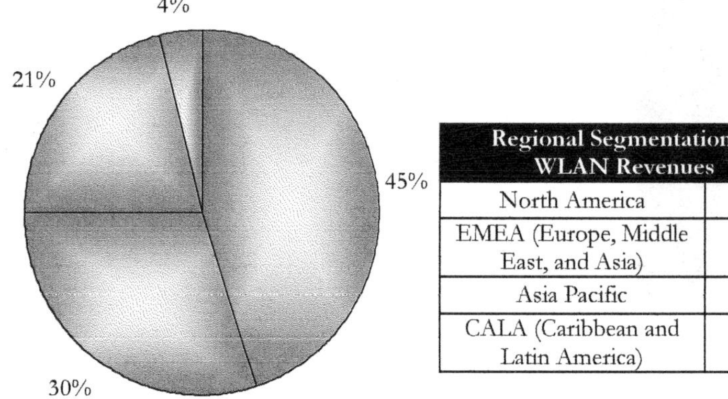


| Regional Segmentation of WLAN Revenues | |
|---|---|
| North America | 45% |
| EMEA (Europe, Middle East, and Asia) | 30% |
| Asia Pacific | 21% |
| CALA (Caribbean and Latin America) | 4% |

# The Players

## Key Players

**Figure 7:** Top Manufacturers of Wireless LAN products[4]

| | Supplier | Website | Distributors | Revenue $M(US) | Market Share % |
|---|---|---|---|---|---|
| 1 | Conexant | www.conexant.com | Avnet; DSD Microtechnology | $220 | 29% |
| 2 | Broadcom | www.broadcom.com | Arrow; Bell Microproducts | $182 | 24% |
| 3 | Texas Instruments (TI) | www.ti.com | Arrow; Avnet; Digi-Key; Newark InOne; Rochester | $91 | 12% |
| 4 | Atheros | www.atheros.com | | $87 | 11% |

| | Supplier | Website | Distributors | Revenue $M (US) | Market Share % |
|---|---|---|---|---|---|
| 5 | Agere | www.agere.com | Almac; Arrow; Echips | $51 | 7% |
| 6 | Marvell | www.marvell.com | Avnet; NuHorizons | $45 | 6% |
| 7 | RF Micro Devices | www.rfmd.com | | $25 | 3% |
| Others | | | | $59 | 8% |
| **Total Revenue** | | | | **$760** | |

## References


1 802.11. (2004). Retrieved Jun. 29, 2005, from Webopedia website: <http://www.webopedia.com/TERM/8/802_11.html>.

2 802.11n WLAN Products to Flourish in 2007. (2005). *Telecom Products*. Retrieved Jun 29, 2005, from <http://www.telecom.globalsources.com /gsol/I/Mobile-wireless/a/9000000060453.htm>.

3 Ahlawat, R., & Canales, C. (2005). 2004 Preliminary Market Statistics: Wireless LAN Equipment Worldwide (G00125841), Retrieved Jun 29, 2005, from Gartner database.

4 Arensman, R. (2004). Wi-Fi winners: Broadcom, TI and Atheros lead booming Wi-Fi chip market. *Electronic Business*, (1097-4881). Retrieved Jun 29, 2005, from TableBase database.

5 Chaplin, K. (2002). Wireless LANs v. wireless WANs. Retrieved Jun. 29, 2005, from Sierra Wireless website: < www.sierrawireless.com/news/docs/2130273_WWAN_v_WLAN.pdf>.

6 Comparing WLAN and WWAN. (2005). Retrieved Jun. 29, 2005, from Wikipedia website: <http://en.wikipedia.org/wiki/Comparing_WLAN_and_LAN.>.

7 Computer Network. (2005). Retrieved Jun. 29, 2005, from Wikipedia website: <http://en.wikipedia.org/wiki/Computer_network>.

8 Harbert, T. (2003, Apr 15). Wanted: more Wi-Fi waves. *Electronic Business*. Retrieved Jun 29, 2005, from <http://www.reed-electronics.com/eb-mag/article/CA290390?pubdate=04%2F15%2F2003>.

9 HomeRF. (2003). Retrieved Jun. 29, 2005, from Webopedia website: <http://www.webopedia.com/TERM/H/HomeRF.html>.

10 Hotspot. (2002). Retrieved Jun. 29, 2005, from Webopedia website: <http://www.webopedia.com/TERM/h/hotspot.html>.

11 IEEE 802.11. (2005). Retrieved Jun. 29, 2005, from Wikipedia website: <http://en.wikipedia.org/wiki/IEEE_802.11>.

12 Introduction to Wireless Networking. (2005). Retrieved Jun. 29, 2005, from Webopedia website: <http://www.tutorial-reports.com/wireless/introduction.php>.

13 Local Area Networks. (2003). Retrieved Jun. 29, 2005, from Webopedia website: < www.webopedia.com/TERM/L/local_area_network_LAN.html>.

14 Webb, R. (2005). Wi-Fi Units Hit Record Volume in 1Q05; Revenue Up 20%. Retrieved from Infonetics website: < www.infonetics.com/resources/purple.shtml?ms05.wl.lq05.nr.shtml>.

15 Wexler, J. (2005). Wireless LAN: State-of-the-Market Report. *Webtutorials*. Retrieved Jun 29,2005, from <http://www.webtorials.com/abstracts/WLAN2005>.

16 Wi-Fi. (2004). Retrieved Jun. 29, 2005, from Webopedia website: <http://www.webopedia.com/TERM/W/Wi_Fi.html>.

17 Wireless Local Area Networks. (2004). Retrieved Jun. 29, 2005, from Webopedia website: <http://www.webopedia.com/TERM/W/WLAN.html>.

# 11

# WWAN

## Wireless Wide Area Network

A **Wireless Wide Area Network** or **WWAN** is a wireless network that spans large areas, such as cities or countries. WANs use satellite systems and antenna sites to make their connections[13]. Wireless WANs differ from wireless LANs because they have a more extensive coverage area and offer connectivity for users, particularly in a business environment. Radio signals over analog, digital cellular, or PCS networks allow the communication of a WWAN to occur.

Several types of cellular systems are used today, the most common currently being 2G TDMA, CDMA, PDC, and GSM. Some 2G and most 3G products are designed for both data and voice transmissions, and the speed of transfer is escalating significantly with each new product. These products can also communicate with the internet using a Wireless Application Protocol, or WAP, which acts as a gateway for information to be sent, processed, received, and returned[3].

**Second Generation** or **2G**, products are analog and digital current networks operating on 800MHz or 1900 MHz spectrums. Advanced Mobile Phone Services (AMPS), TDMA, GSM, and CDMA are all included in the 2G product range[3].

**Third Generation** or **3G**, also called Universal Mobile Telecommunications Systems (UMTS), are standards that enhance services, such as more extensive roaming capabilities and higher transfer rates[3].

**Time Division Multiple Access** or **TDMA** is a form of WWAN that delivers digital wireless service using Time-Division Multiplexing (TDM)[14]. TDMA splits the radio frequency into time slots and then allocates each slot to phone calls. With this technology, a single frequency can support multiple and simultaneous data channels. GSM and other wireless products incorporate TDMA into their technology[14].

**Code Division Multiple Access** or **CDMA** is the other top digital cellular technology. CDMA differs from GSM[1] in that it does not reserve a specific frequency exclusive to the user but provides access to the full available spectrum. According to Qualcomm, CDMA was selected in 1999 by the International Telecommunications Union (ITU) as the industry standard for new third generation wireless systems[6]. It is still widely used today. ITU is preparing for a new deployment of technology in 2005. Qualcomm, the patent holder of CDMA, has developed a third generation technology that will be more effective at converging technologies in mobile phones with digital cameras, video camcorders, MP3 players, and handheld computers[6].

**Personal Digital Cellular** or **PDC** is a 2G TDMA-based protocol that is used in Japan and operates in the 800 and 1500 MHz bands[3].

**Global System for Mobile Communication** or **GSM** is a software system that is not controlled by any one company, meaning it is open or non-proprietary. It is a constantly evolving software that boasts its international roaming capabilities and even has extended service where terrestrial coverage isn't available[4]. GSM is different from other first generation wireless systems because it uses TDMA. Each spoken word is digitally encoded and thus it can create the characteristics of human speech. GSM's ability to evolve has enabled it to house other technologies such as GPRS (General Packet Radio Service), which is a packet-based wireless WAN that uses cellular networks to transmit data in packet form. This type of wireless communication opens the door for internet, email, and other non-voice applications to cell phones and PDAs that have GPRS chipsets[7].

**Wideband Code Division Multiple Access (WCDMA)** is a third generation or 3G, technology that uses the CDMA air interface rather than TDMA in GSM systems[16]. Completed in 1999, it will be one of the main technologies for 3G cellular networks. WCDMA is much more complicated and "compute-intensive" than what is in 2G cellular systems; consequently it will be a difficult product to integrate[17].

# The Product

## Applications and End Uses

The majority of WWAN usage is in cellular networks, PDAs, handsets, and Smartphones. This allows business travelers, for example, to connect to their office network, email, and internet while away[2]. Also, WWAN technology is used in transportation communication. Specifically, it is used to get information to drivers, such as providing directions to destinations and bills of lading[5].

## Product Features and User Benefits

**Figure 1:** Product Features and User Benefits[2,10,11]

| Product Features | User Benefits |
|---|---|
| 9.6k Data Transfer Rate | Cellular phone is able to send and receive data from the internet. GPRS allows the data transfer rate to be even faster. |
| Security Authentication | Network operators can verify the identity of a subscriber, which can help recover lost as well as stolen phones. |
| Security Confidentiality | Voice, data, and sensitive signalling information can be protected against eavesdropping or tapping on the radio path. |
| Anonymity in Security | Tracking user location or calls made or received from user by tapping or eavesdropping is prevented. |
| No Physical Constraints | Requires no cables or excess hardware. |
| Coverage | Allows for mobility, a broader availability of connectivity, and fast access to information. Provides nationwide and global coverage as well. |
| Transfer Speeds from 5 kbps - 20 kbps | Faster than dial-up; however, not yet faster than a WLAN. |

## Life Cycle

Wireless Wide Area Networks are constantly developing. New products are coming out frequently while old products continue to evolve. The Life Cycle Curve in Figure 2 illustrates the older products, beginning with CDMA, and ending with the newest products. As seen in the curve, as soon as one product reaches the growth stage, another is introduced, and yet another is in the research and development stages of its life.

**Figure 2:** Life Cycle Curve for WWAN technologies

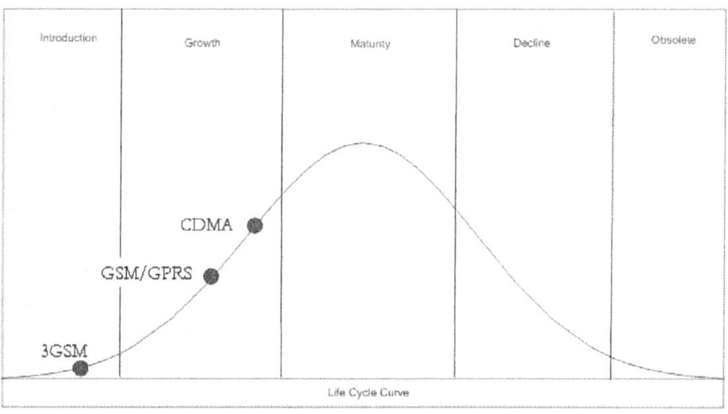


# The Market

## Market Trends

**Figure 3:** Global Digital Mobile Market[8]

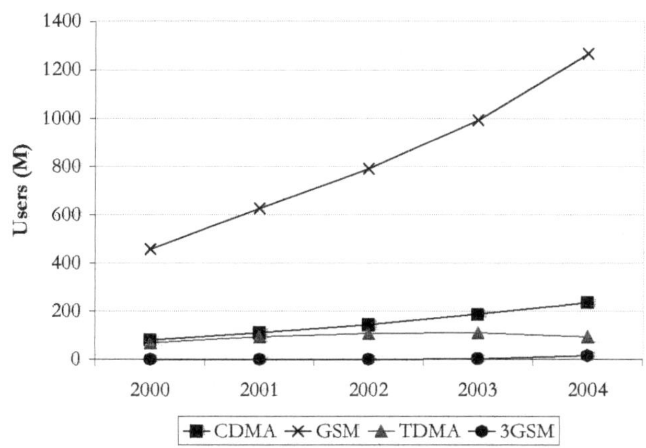


| | 2000 | 2001 | 2002 | 2003 | 2004 |
|---|---|---|---|---|---|
| CDMA | 80.3 | 110.9 | 144.1 | 186.7 | 236.3 |
| GSM | 456.1 | 626.2 | 790.6 | 991.7 | 1266.4 |
| TDMA | 67.6 | 94.1 | 108.1 | 109.1 | 93.7 |
| 3GSM | 0 | 0 | 0.2 | 2.7 | 16.2 |

In 2004, less than ten years after its release, GSM was the world's most popular and fastest growing mobile standard, spanning 200 countries. One in every six people across the world used GSM in 2004; in January 2004, there were over 1 billion users[15].

The future is bright for GSM as EDGE (Enhanced Data Rates for Global Evolution) brings up to 384 Kilobits per second (48 Kilobytes) and has or will be launched globally[7]. To get the pulse of the cellular chipset market, study statistics on GSM growth, which has been the largest user of cellular chipsets since its inception. GSM technologies that include GPRS and EDGE (Enhanced Data Rates for Global Evolution) will still dominate the market in 2008, comprising a forecasted 57% of the total market[18].

CDMA will make up 20% of the market while WCDMA will comprise 21%. WCDMA is forecasted to be the fastest growing technology and is projected to have a Compound Annual Growth Rate, CAGR, of 82% between 2004 and 2008. EDGE and the newest CDMA technology will account for the second and third fastest growing technologies[18].

## Regional Market Trends

**Figure 4:** GSM Growth by Region (2004)[8]

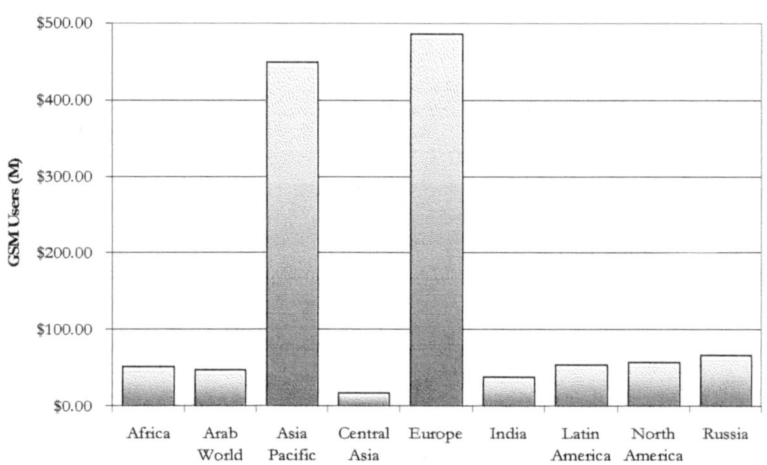


| Regional GSM Revenues, 2004 | | | |
|---|---|---|---|
| Africa | 51.2 | India | 37.7 |
| Arab World | 47.0 | Latin America | 53.7 |
| Asia Pacific | 448.8 | North America | 57.3 |
| Central Asia | 17.8 | Russia | 66.8 |
| Europe | 486.1 | **Global GSM Total** | **1266.4** |

## Market Segmentation

The WWAN market, according to iSuppli Corporation, is primarily made up of mobile handsets. Figure 5 shows the billions of dollars spent each year on WWAN, used for mobile handsets.

**Figure 5:** Market Segmentation for WWAN products, Mobile Handsets

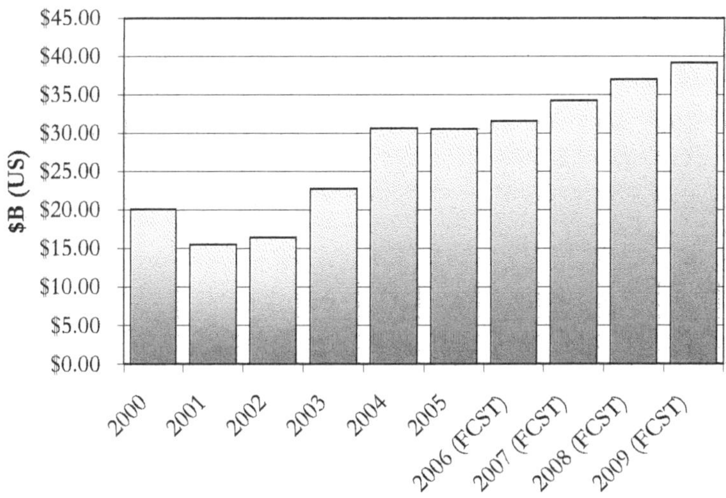


| 2000 | 2001 | 2002 | 2003 | 2004 | 2005 |
|---|---|---|---|---|---|
| $20.1 | $15.5 | $16.4 | $22.8 | $30.7 | $30.6 |

| 2006 (FCST) | 2007 (FCST) | 2008 (FCST) | 2009 (FCST) |
|---|---|---|---|
| $31.6 | $34.2 | $37.0 | $39.1 |

Source: iSuppli Corporation

# The Players

## Key Players

**Figure 6:** Top Manufacturers of WWAN Products in 2002[12]

| | Supplier | Website | Distributors | Market Share % |
|---|---|---|---|---|
| 1 | Texas Instruments | www.ti.com | Arrow; Avnet; Digi-Key; Newark InOne | 9.3% |
| 2 | Qualcomm | www.qualcomm.com | | 8.87% |
| 3 | Freescale (Motorola) | www.freescale.com | Arrow; Avnet; Digi-Key; Future; Newark InOne; Richardson; Rochester | 7.69% |
| 4 | ST Microelectronics | www.st.com | Arrow; Avnet; Digi-Key; Future; Mouser NuHorizons | 7.54% |
| 5 | Intel | www.intel.com | Arrow; Avnet; Cilicon; Digi-Key; Rochester | 6.18% |

| | Supplier | Website | Distributors | Market Share % |
|---|---|---|---|---|
| 6 | Philips Semiconductors | www.semiconductors.philips.com | All American; Avnet; Classic Components | 6.08% |
| 7 | Infineon Technologies | www.infineon.com | ACG North America; Arrow; Avnet; Digi-Key; Future | 4.42% |
| 8 | NEC Electronics | www.nec.com | Arrow; Avnet; Bell Microproducts; Edge; Future; Symmetry | 3.52% |
| 9 | Toshiba | www.toshiba.com | Arrow; Avnet; NuHorizons; TAEC | 3.42% |
| 10 | Agere | www.agere.com | Almac; Arrow; Avnet; Echips | 3.07% |

## References


1 CDMA. (2005). Retrieved Jun. 29, 2005, from Webopedia website: <http://www.webopedia.com/TERM/C/CDMA.html>.

2 Chaplin, K. (2002). Wireless LANs v. Wireless WANs. Retrieved Jun. 29, 2005, from Sierra Wireless website: < www.sierrawireless.com/news/docs/2130273_WWAN_v_WLAN.pdf>.

3 Fisher, J., & Wang, R. (2005). Wireless wide area networks (WWANs). *pdaMD*, Retrieved Jun 29, 2005 from; <www.pdamd.com/vertical/features/ >.

4 Frequently Asked Questions. (2005). Retrieved Jun. 29, 2005, from GSM World website: <http://www.gsmworld.com/technology/faq.shtml>.

5 Gandavabi, B. (2005). The power of fusion. Retrieved Jun. 29, 2005, from Industrial Data & Information, Inc. website: <www.idii.com/wp/pmtWlanWwan.pdf>.

6 Glossary. (2005). Retrieved Jun. 29, 2005, from Qualcomm CDMA Technologies website: <http://www.cdmatech.com/resources/glossary.jsp>.

7 GPRS Glossary and Definitions. (2005). Retrieved Jun. 29, 2005, from t-mobile website: <http://www.support.t-mobile.com/knowbase/root/public/tm21389.ht>.

8 Growth of the Global Digital Mobile Market. (2005). Retrieved Jun. 30, 2005, from GSM Association website: <http://www.gsmworld.com/news/statistics/pdf/gsma_stats_q4_04.pdf>.

9 GSM Facts and Figures. (2005). Retrieved Jun. 30, 2005, from GSM World website: <http://www.gsmworld.com/news/statistics/index.shtml>.

10 GSM - The Wireless Evolution. (2005). Retrieved Jun. 30, 2005, from GSM World website: <http://www.gsmworld.com/technology/index.shtml>.

11 Howard, P. (2004). GSM and UMTS Security. Retrieved Jun. 30, 2005, from Royal Holloway, University of London website: <http://www.isg.rhul.ac.uk/msc/teaching/ic3/IC30405pt7.ppt#68>.

12 McClean, B. (2002). Top Ten Wireless Communications IC Suppliers. Retrieved Jun. 30, 2005, from IC Insights website: <http://www.icinsights.com/news/releases/press20021204.pdf>.

13 Tabona, A. Z. (2004, May 20). An Introduction to Wireless Networking (Part 1). Retrieved Jun 29, 2005, from Windows Networking website: <http://www.windowsnetworking.com/pages/article_p.asp?id=310>.

14 TDMA. (2003). Retrieved Jun. 29, 2005, from Webopedia website: <http://www.webopedia.com/TERM/T/TDMA.html>.

15 Today's GSM Platform. (2005). Retrieved Jun. 30, 2005, from GSM World website: <http://www.gsmworld.com/technology/gsm.shtml>.

16 WCDMA. (2005). Retrieved Jun. 29, 2005, from Webopedia website: <http://www.webopedia.com/TERM/W/WCDMA.html>.

17 WCDMA (UMTS). (2003). Retrieved Jun. 29, 2005, from UMTS World website: <http://www.umtsworld.com/technology/wcdma.htm>.

18 Wood, B., Milanesi, C., Liang, A., De La Vergne, H., Mitsuyama, N. (2004). Forecast: Mobile Terminals, Worldwide, 1999-2008 (Executive Summary) (G00124055), Retrieved Jun 30, 2005, from Gartner database.

# Logic

# 12 Discrete Logic

Logic is the domain of human reasoning that tells you a certain proposition is true if certain conditions are true. **Logic**, as applied to digital circuits, implements logical functions. It is a science dealing with the basic principles and applications of truth tables and Boolean algebra. The three basic logic symbols are AND, OR, and NOR. When used in Boolean algebra, these symbols are very similar to addition, subtraction, and multiplication. The digital logical circuits are the basic elements that form the building blocks for complex digital systems[6].

The three basic logic elements (AND, OR, and NOR) can be combined to form more complex logic circuits and complete digital systems. Common logic functions are comparison, arithmetic, code conversation, encoding, decoding, data selection (multiplexer, demultiplexer), storage (flip flop, register, semiconductor, memories, magnet memories), and counting[6].

If you open the backplane of a computer and remove a board from the rack, you will find the motherboard. It has a collection of chips connected by trails of metal along the board. All values inside the computer are stored as 0s and 1s, and they are transmitted along these metal trails using one voltage to represent a zero, another voltage to represent a one. Inside the IC package are devices called Logic Gates. These **logic gates** are small (several micron) structures that take one or more bits as input, and produce another bit as output, manipulating the 0s and 1s. The 0s and 1s can be transmitted at the speed of electrons flowing through the chip, and complex calculations are quickly completed by using an entire sequence of these tiny gates[6].

Logic gates serve as the building blocks to digital logic circuits, using combinational logic. Most logic gates have two inputs and one output. At any given moment, every terminal is in one of the two binary conditions, low (0) or high (1), represented by different voltage levels. The logic state of a terminal can, and generally does, change often, as the circuit processes data[6]. The most common logic gates used perform these logic functions:

- **AND**: output is 1 if all inputs are 1
- **OR**: output is 0 if all inputs are 0
- **XOR**: output is the sum of all inputs
- **NOT**: output is the opposite of the single input
- **NAND**: output is 0 if all inputs are 1
- **NOR**: output is 1 if all inputs are 0
- **XNOR**: output is the inverse of the XOR gate

By linking the output of one gate to the inputs of other gates, a more powerful calculation can be carried out. The calculations and functions are listed and described in Figure 1.

**Figure 1:** Calculations Using Gates

| Calculation/Function | Description |
|---|---|
| **Binary Adders** and **Parallel Binary Adders** | Logic devices that add two binary numbers to form a sum with optional carry. |
| **Digital Dividers** | Digital circuits that produce the arithmetic operation of binary division. |
| **Comparators** | A logic function that performs magnitude comparisons. It compares two quantities and indicates whether they are equal or not. |
| **Decoders** and **Encoders** | Electronic devices that separate or combine two or more signals; code converters. |
| **Inverters** | The logic circuit that performs the NOT operation (changes one logic level to the opposite logic level). |
| **Multiplexers** and **Demultiplexers** | A logic circuit that switches digital data from several input lines onto a single output line in a specified time sequence or visa versa. |

A **flip flop** is a basic storage circuit that can store only one bit at a time. Its output can first flip one way and then flop back the other way. It maintains its state until a trigger impulse forces it to its alternate state. This alternate state is held until another signal is received, which forces it back to its original state[3]. Figure 2 describes a number of different types of flip flops.

**Figure 2:** Types of Flip Flops[3]

| Type | Description |
|---|---|
| **Latches** | Bistable digital circuits used for storing a bit. They are sets of Flip Flops with a common clock signal. In each cycle, they take the logical input signals over to their outputs. |
| **Edge Triggered Flip Flop** | A type of digital storage circuit in which the entered data appears on the output on the same clock edge. |
| **Master Slave Flip Flop** | A type of digital storage circuit in which the entered data appears on the output on the same clock edge. |
| **Registers** | Collections of Flip Flops that move stored bits in one or both directions, usually serially. They are digital circuits capable of storing and shifting binary information and typically used as temporary storage devices. |

The basic purpose of counters and timers is to count events represented by changing levels or pulses. A **counter** is a digital circuit capable of counting electronic events by progressing through a sequence of binary states. A **timer** is a circuit that can be used as a one shot or as an oscillator. They both have a clock input and a number of count outputs, which give the number of clock cycles. The outputs may change on either rising or falling clock edges. These edges are known as **TON** (Timer On Delay) and **TOF** (Timer Off Delay) for timers; and **CTD** (Count Down) and **CTU** (Count Up) for counters.

An **asynchronous counter** is a type of counter in which each stage is clocked from the output of the preceding stage. A **synchronous counter** is a type of counter in which the same pulse clocks each stage. **Cascaded counters** and **cascaded timers** connect "end-to-end" as when several counters are connected from the terminal count output of one counter to the enable input of the next counter.

**Discrete logic** is an individual component or group of components on a chip, used to perform one function or set of functions. Although up to hundreds of thousands of gates are routinely placed on a single chip, discrete logic chips contain one or two gates and are used as glue logic chips to transfer or reshape signals between integrated circuits (ICs) and application specific integrated circuits (ASIC). **ASIC**s are the small chips linking the brain chips together, allowing them to exchange information[4]. Figure 3 breaks down the major technologies and families that make up discrete logic integrated circuits.

**Figure 3:** Technologies and Families of Discrete Logic

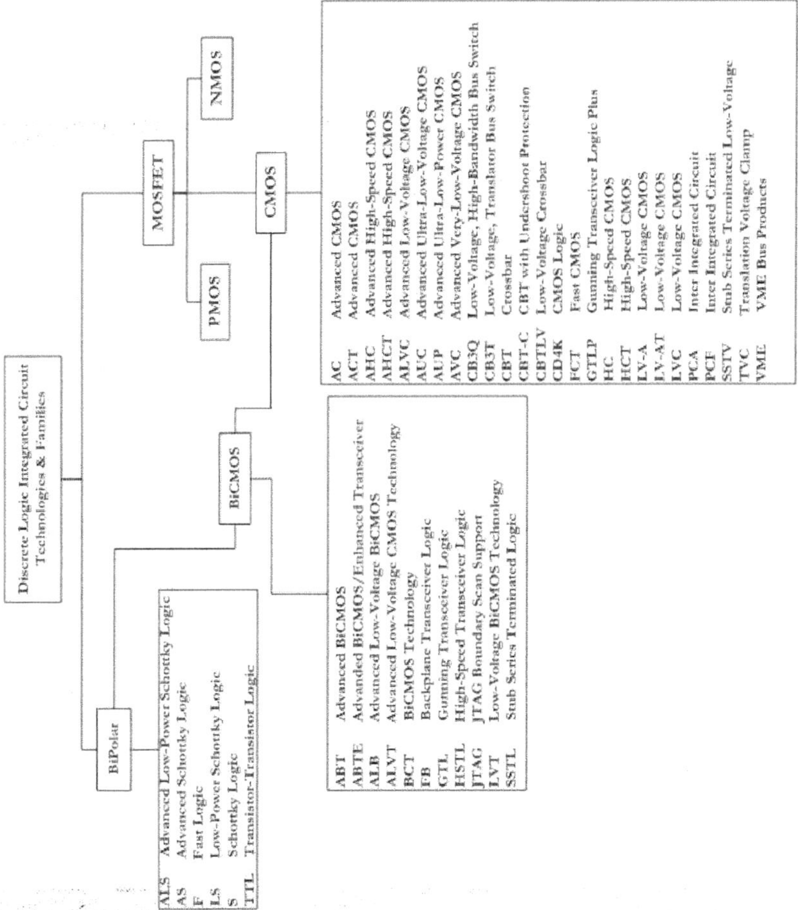


Electronic logic gates can be produced using different production technologies. In industry, integrated circuits with logic gates, produced using production technology, belong to a certain logic family of devices. The list of logic families can be divided into key technologies with further subdivision into type, as seen in the chart above. Each subdivision is referred to as the **logic product family**.

Discrete logic is the glue used to interconnect the larger more complex chips. These simple "building block" logic gate integrated circuits are the reason for the various logic families; however, most families are based on bipolar or the Transistor-Transistor Logic (TTL), CMOS, and BiCMOS-logic families, each is described below.

The **bipolar logic circuits** were the first developed from discrete components, components independent of others. Included in this group are Resistor-Transistor Logic (RTL), Diode-Transistor Logic (DTL), Transistor-Transistor Logic (TTL), and Emitter-Coupled Logic (ECL). The first microelectronic technology, Resistor-Transistor Logic (RTL), used mostly transistors and only a few resistors. In Diode-Transistor Logic (DRL), performance was improved by substituting semiconductor diodes for many of the resistors. Transistor-Transistor Logic (TTL), in which transistors are abundant and coupled directly, remains the most popular bipolar technology today. In the TTL class, digital circuits are built from bipolar junction transistors and resistors. It is known for being a widespread integrated circuit family used in many applications such as computers, industrial controls, and measuring instruments[5]

In 1962, Texas Instruments introduced the 7400 series of ICs, which had a wide range of digital logic block functions. This made TTL popular with electronic systems designers. The Texas Instrument 7400 series became an industry standard; however, other TTL devices were made by Motorola, Signetics (now Phillips), National Semiconductor, and many other companies. TTL became important because for the first time, low-cost Integrated Circuits (IC) made digital techniques economically practical for tasks previously done by analog methods. Each chip can contain the equivalent of a few hundred transistors, and each of these integrated circuits can perform a separate function such as Logic Gates, Flip Flops, Latching, Counters, Adders, Timers, Read-Write and Read-Only Memory, Level Converters, Multifunction Logic, and so on. TTL devices have become important for glue logic applications. TTL does have some disadvantages that include dissipating a lot of energy and slow speed, compared to other current devices[5]

**Figure 4:** The Bipolar Family (variations of the original TTL logic)[10]

| Bipolar | |
|---|---|
| ALS | Advanced Low-Power Schottky Logic |
| AS | Advanced Schottky Logic |
| F | Fast Logic |
| LS | Low-Power Schottky Logic |
| S | Schottky Logic |
| TTL | Transistor-Transistor Logic |

Another group of circuits is **Metal Oxide Silicon Field Effect Transistors** (MOSFET, also known as FET) circuits. MOSFET technologies reduced the space issue once again and also drew less power than the bipolar devices. The first MOSFET electronic circuits employed p-channel (PMOS) devices because they were the easiest to make. As MOSFET technology advanced, n-channel (NMOS) devices replaced PMOS devices because they offered higher speed performance for the same density, complexity, and cost. The need for reduced power consumption led to the development of the larger but more power-efficient, complementary devices, known as CMOS[8].

The **CMOS** family is a major class of integrated logic circuit using the MOSFET technology. It is made from a complementary pair of field effect transistors, a NMOS and PMOS, resulting in the name CMOS, complementary metal-oxide-semiconductor. It was introduced because a low-power alternative was in demand. A very important feature of CMOS chips is that they work with a broader range of power supply voltages. Lowering the supply voltage reduces the current required to charge stray capacitance, and so reduces the current drawn by complex microprocessors. This in turn reduces the heat dissipation of the processor. Over the years, many logic families emerged that used the CMOS technology, and improvements were made trying create a logic device that combined high speed with low power dissipation and that was compatible with older logic families[1]. Figure 5 shows most of these families[10].

**Figure 5:** CMOS Families

| CMOS | |
|---|---|
| AC | Advanced CMOS Logic |
| ACT | Advanced CMOS Logic |
| AHC | Advanced High-Speed CMOS |
| AHCT | Advanced High-Speed CMOS |
| ALVC | Advanced Low-Voltage CMOS Technology |
| AUC | Advanced Ultra-Low-Voltage CMOS Logic |
| AUP | Advanced Ultra-Low-Power CMOS Logic |
| AVC | Advanced Very-Low-Voltage CMOS Logic |
| CB3Q | Low-Voltage, High-Bandwidth Bus Switch Technology |
| CB3T | Low-Voltage, Translator Bus Switch Technology |
| CBT | Crossbar Technology |
| CBT-C | CBT with Undershoot Protection |
| CBTLV | Low-Voltage Crossbar Technology |
| CD4000 | CMOS Logic |
| FCT | Fast CMOS Technology |
| GTLP | Gunning Transceiver Logic Plus |
| HC | High-Speed CMOS Logic |
| HCT | High-Speed CMOS Logic |

| LV-A | Low-Voltage CMOS Technology |
|---|---|
| LV-AT | Low-Voltage CMOS Technology |
| LVC | Low-Voltage CMOS Technology |
| PCA | Inter Integrated Circuit |
| PCF | Inter Integrated Circuit |
| SSTV | Stub Series Terminated Low-Voltage Logic |
| TVC | Translation Voltage Clamp |
| VME | VME Bus Products |

BiCMOS logic is a logic device that combines CMOS inputs and TTL drivers. The BiCMOS family has many members, shown in Figure 6[10].

**Figure 6:** BiCMOS Family Members

| BiCMOS | |
|---|---|
| ABT | Advanced BiCMOS Technology |
| ABTE | Advanced BiCMOS Technology / Enhanced Transceiver Logic |
| ALB | Advanced Low-Voltage BiCMOS |
| ALVT | Advanced Low-Voltage CMOS Technology |
| BCT | BiCMOS Technology |
| FB | Backplane Transceiver Logic |
| GTL | Gunning Transceiver Logic |
| HSTL | High-Speed Transceiver Logic |
| JTAG | JTAG Boundary Scan Support |
| LVT | Low-Voltage BiCMOS Technology |
| SSTL | Stub Series Terminated Logic |

# The Product

## Application and End Uses

Logic circuits are found in most all electronic equipment. However, since FPGAs and ASICs have been developed, these devices have replaced a number of logic circuits because programmable logic devices enable the engineer to combine functions that were previously done using individual discrete devices. This has caused a decline in the worldwide volume of discrete logic circuits. The primary users of discrete logic circuits include: wireless, telephones, and computers, while instrumentation, medical technology, and military application comprise the secondary markets for discrete logic.

Listed below are some typical applications for discrete logic circuits. However, they are not limited to these.

- Digital Still Cameras
- Digital Camcorders
- Videogame Consoles
- Smart Handheld Devices
- Information/Internet Appliances
- Compressed Audio Players
- DVD Players
- DVD Recorders
- Personal Video Recorders
- Digital Cable Set-Top Boxes
- Digital Satellite Set-Top Boxes
- Integrated Digital Television Sets
- High Definition Television
- Digital Televisions
- Telephones
- Telephones with Digital Cameras
- Flat-Screen Television and Monitors

## Product Features and User Benefits

In today's society, features are extremely high tech. The features of discrete logic range from handling a high-voltage potential to dynamic output control circuitry that provides high-speed and low-noise for future PCs and workstations.

**Figure 7:** Product Features and User Benefits of Discrete Logic

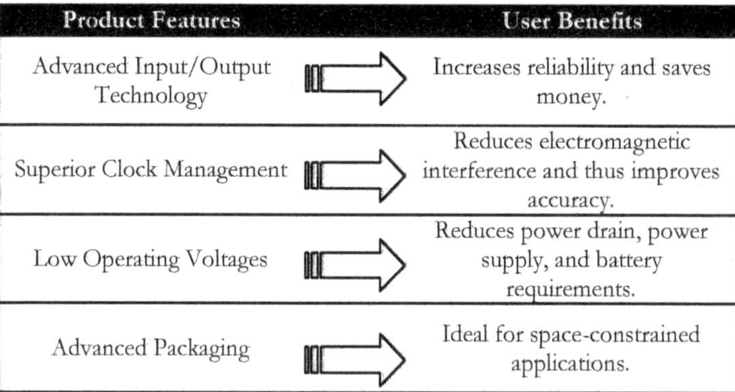

| Product Features | User Benefits |
|---|---|
| Advanced Input/Output Technology | Increases reliability and saves money. |
| Superior Clock Management | Reduces electromagnetic interference and thus improves accuracy. |
| Low Operating Voltages | Reduces power drain, power supply, and battery requirements. |
| Advanced Packaging | Ideal for space-constrained applications. |

## Life Cycle Curve

Computers have evolved and become what they are today due mainly to the evolution of Boolean gates and discrete logic. The first gates used relays, which were big and slow. Vacuum tubes replaced relays. The tubes were faster but still too big. They also burnt out much like a light bulb. In 1947, the transistor was perfected. This revolutionized discrete logic and computer technology. The transistor was more reliable, smaller, and more power-efficient than the tubes[7]. In the 1960s, researchers came up with ICs using discrete logic coupled with the transistor.

**Figure 8:** Life Cycle Curve for Discrete Logic[10]

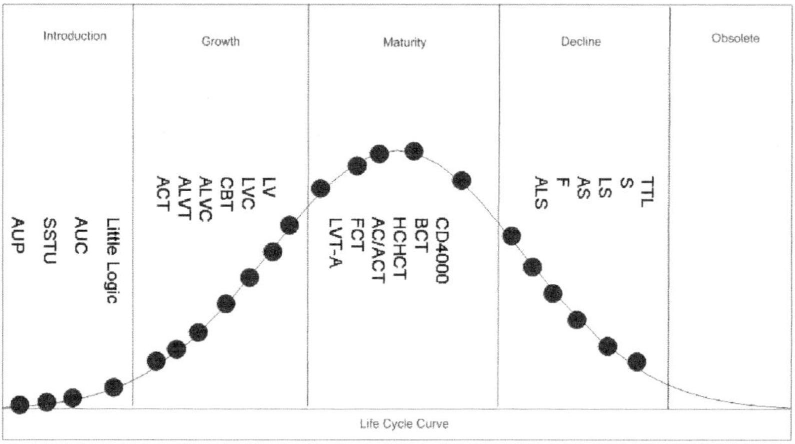

# The Market

## Total Available Market

Although, discrete logic only makes up 3.7% of the total semiconductor market, the expected total available market for discrete logic within consumer electronics in 2005 is to be $3.5 billion (U.S.). The total expected discrete-semiconductor market is currently 14 billion, and has a growth rate of 10% per year (ir.vishay.com). On Semiconductor claims that you can find at least 25 discrete- and logic-IC components in any cellular phone (www.edn.com). These devices are in almost every electronic device in your home, office, or car. The companies that sell these can only make their profits on extremely high volume. However, with the increased improvements and use of programmable logic ICs, the discrete logic market seems to be leveling out.

**Figure 9:** Worldwide Semiconductor Device Forecast 2000-2006[2]

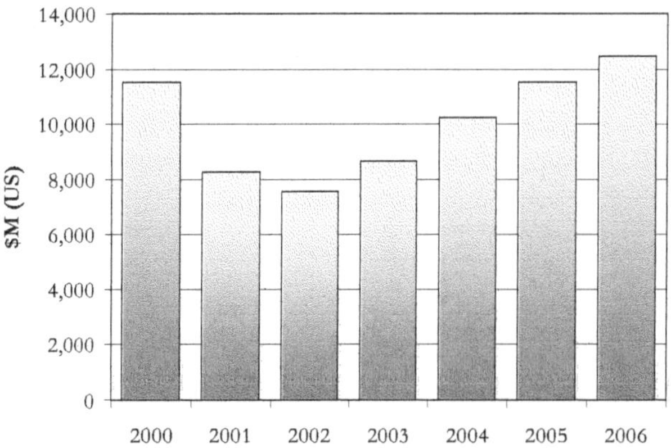


| 2000 | 2001 | 2002 | 2003 | 2004 | 2005 (FCST) | 2006 (FCST) |
|---|---|---|---|---|---|---|
| $11,540 | $8,263 | $7,561 | $8,686 | $10,267 | $11,532 | $12,476 |

## Market Segmentation

**Figure 10:** Logic Use in Consumer Electronics[2]

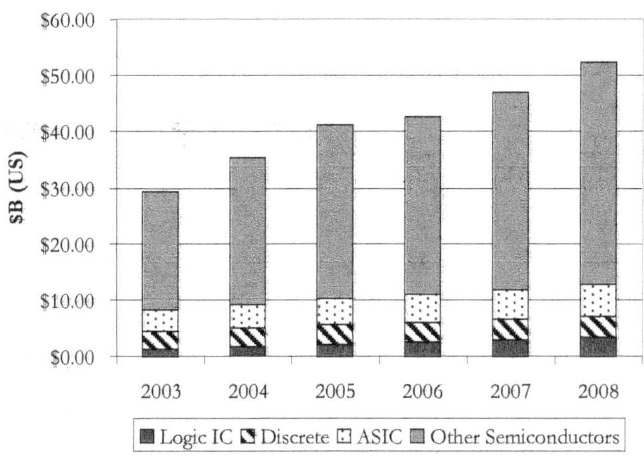


| | 2003 | 2004 | 2005 | 2006 | 2007 | 2008 |
|---|---|---|---|---|---|---|
| **Logic IC** | $1.4 | $1.7 | $2.2 | $2.5 | $2.9 | $3.4 |
| **Discrete** | $3.1 | $3.4 | $3.5 | $3.6 | $3.7 | $3.7 |
| **ASIC** | $3.9 | $4.2 | $4.7 | $4.9 | $5.3 | $5.7 |
| **Total Semiconductor Sales** | **$29.4** | **$35.5** | **$41.2** | **$42.6** | **$47** | **$52.4** |

# The Players

## Key Players

**Figure 11:** Top 10 Discrete Logic Manufacturers in 2003[9]

| | Supplier | Website | Distributors | Revenue $M (US) | Market Share % |
|---|---|---|---|---|---|
| 1 | Texas Instruments (TI) | www.ti.com | Arrow; Avnet; Digi-Key; Newark InOne | $360 | 26.5% |
| 2 | Toshiba | www.toshiba.com | Arrow; Avnet; NuHorizons, TAEC | $277 | 20.4% |
| 3 | Fairchild Semiconductor | www.fairchildsemi.com | All American; Arrow; Avnet; Bell Components; Future | $178 | 13.1% |
| 4 | Philips Semiconductors | www.semiconductors.philips.com | Arrow; Avnet; Digi-Key; Future | $164 | 12.1% |
| 5 | ON Semiconductor | www.onsemi.com | All American; Arrow; Avnet; Chip Supply; Digi-Key; Future; FAI; Minco; Newark InOne; Rochester; Semi Dice | $131 | 9.65% |

| | Supplier | Website | Distributors | Revenue $M (US) | Market Share % |
|---|---|---|---|---|---|
| 6 | ST Microelectronics | www.st.com | Arrow; Avnet; Future; NuHorizons | $95 | 7% |
| 7 | IDT | www.idt.com | Avnet; Arrow; Earle Associates; Integrated Device Technology; Norpcomp; SJ; Victory Sales; Universal Technology; Port Electronics | $56 | 4.1% |
| 8 | Renesas | www.renesas.com | All American; Arrow; Avnet; Digi-Key; NuHorizons | $44 | 3.2% |
| 9 | National Semiconductor (NSC) | www.national.com | Arrow; Avnet; Future; Digi-Key; Semi Dice | $31 | 2.28% |
| 10 | Intersil | www.intersil.com | Arrow: Avnet; Digi-Key; Newark InOne; NuHorizons; | $21 | 1.55% |

## References

1 CMOS. Retrieved on Jun.03, 2005, from wikipedia website: <http://www.wikipedia.com>.

2 Carbone, J. (2004, Sept 16). Expect price erosion for CE chips. *Purchasing Magazine Online*. Retrieved Jun 06, 2005, from <http://www.purchasing.com/index.asp?layout=articlePrint&articleID=CA454034>.

3 Digital Logic. Retrieved on Jun.03, 2005, from Williamson Labs website: <http://www.williamson-labs.com/>.

4 Discrete Logic. Retrieved on Jun.03, 2005, from Techweb website: <http://wwwl.techweb.com/>.

5 Electronic Gates. Retrieved on Jun.03, 2005, from wikipedia website: <http://www.wikipedia.com>.

6 Floyd, T. (2002). *Digital Fundamentals*. New Jersey: Prentice Hall.

7 How Semiconductors Work. Retrieved Jun. 03, 2005, from How Stuff Works website: <http://www.howstuffworks.com>.

8 LaPedus, M. (2004, Nov 22). Samsung, Infineon gain ground in top-10 chip rankings (courtesy of silicon strategies). *My-ESM.com.*, Retrieved Jun 06, 2005, from <http://www.my-esm.com/showArticle.jhtml?articleID=54200162>.

9 Lawrence, Rowe and Patricia Schank. An Introduction to Microelectronics Manufacturing and Markets. Retrieved on Jun.06, 2005, from Berkeley University website: <http://www.bmrc.berkeley.edu/courseware/ICMfg92/>.

10Logic. Retrieved Jun.03, 2005, from Texas Instruments website: <http://focus.ti.com/logic/docs/logicportal.tsp?templateId=5985>

# 13 Programmable Logic

A **Programmable Logic Device** or **PLD** is an integrated circuit that is used to perform digital logic functions in a system utilizing fewer ICs and enabling greater flexibility. Unlike a logic gate, which has a fixed function, a PLD is not programmed at the time of manufacturing unless it is an application specific integrated circuit (ASIC). PLDs are standard, off-the-shelf, integrated circuits that the user or a third-party provider can program to perform certain functions[11]. They can contain hundreds to thousands of logic gates that must be programmed before being used in a system. Programmable logic devices offer customers a wide range of logic capacity, features, speed, and voltage characteristics, and these devices range from one-time programmable to field programmable in order to perform a number of functions. PLDs are ideal for control applications[12].

The memory is used to store the pattern that was given to the chip during programming. Most of the technologies for storing data in an integrated circuit have been adapted for use in PLDs. The most common PLD technology is known as **fusible-link**, or **fuse**—a technology that contains tiny fuses that may be programmed and configured to perform almost any logical operation. It is a one-time programmable (OTP) device, meaning that once it has been programmed, it cannot be changed. **Antifuse** technology is an alternative to fusible-link, by which a programming unit is used to apply voltage, resulting in a connection pattern designed to perform the desired function. This technology is also OTP. **Reprogrammable PLDs** are PLDs in which each fuse of an OTP device is replaced with a static RAM, or SRAM, cell, allowing the device to be reprogrammed while remaining on the circuit board[13]. The field programmable gate array provides the reprogrammability.

The list below consists of most of the technologies that are used in FPGAs:

- SRAM: Based on static memory technology. In-system programmable and re-programmable. Requires external boot devices. Usually CMOS.
- Anti-fuse: One-time programmable. CMOS.
- EPROM: Electrically Programmable Read-Only Memory technology. Usually one-time programmable in production because of plastic packaging. Windowed devices can be erased with ultraviolet (UV) light. CMOS.

- EEPROM: Electrically Erasable Programmable Read-Only Memory technology. Can be erased, even in plastic packages. Some, but not all, EEPROM devices can be in-system programmed. CMOS.
- FLASH: Flash-erase EPROM technology. Can be erased, even in plastic packages. Some, but not all, FLASH devices can be in-system programmed. Usually, a FLASH cell is smaller than an equivalent EEPROM cell and is therefore less expensive to manufacture. CMOS.
- Fuse: One-time programmable. Bipolar.

The first programmable logic devices are known as simple PLDs (SPLD). They were introduced to commercial market under the name of programmable array logic (PAL)™. An improvement of the PAL was the generic array logic device (GAL). This device has the same logical properties as the PAL but can be erased and reprogrammed. PALs and GALs are available only in small sizes, of a few hundred logic gates. For bigger and more complex logic circuits, complex PLDs or CPLDs must be used. They contain several PALs sharing the same programmable interconnection. CPLDs can replace hundreds of thousands of logic gates[12].

**Field Programmable Gate Arrays (FPGAs)** are devices based on gate array technology. It is a gate array that can be reprogrammed after it is manufactured, rather than having its programming fixed during the manufacturing. It has the highest amount of logic density, the most features, and the highest performance. FPGAs consist of an array of programmable logic blocks, similar to an ordinary gate array, that contain logic gates and memory elements. These blocks can be configured to provide a variety of functions. They are connected by programmable row and column channels surrounded by programmable input/output blocks. Some FPGAs are OTP, where others are SRAM-based and may be reprogrammed. FPGAs are used in a large variety of applications, ranging from data processing and storage, to instrumentation, telecommunications, and digital signal processing. FPGAs and CPLDs are often equally good choices for a particular task and the decision is often more economic than technical[7].

FPGA device manufacturers include Xilinx, Altera, Lattice Semiconductor, Actel, Cypress, Atmel, QuickLogic, and ST Microelectronics; the devices are used in applications consisting of DSP, software-defined radio, aerospace and defense systems, ASIC prototyping, medical imaging, computer vision, speech recognition, cryptography, bioinformatics, and a growing range of other areas.

A chip that is custom designed for a special application, rather than a general purpose such as a microprocessor, is called an **Application Specific**

™ Programmable Array Logic (PAL) is originally trademarked by MMI, who was bought in 1987 by AMD who created a consolidated operation called Vantis, which was finally acquired by Lattice Semiconductor in 1999.

**Integrated Circuit (ASIC)**. It performs a fixed set of steps that cannot be changed. The use of an ASIC improves performance over general purpose CPUs because they are constructed to do a specific job. An ASIC is typically faster, smaller, and cheaper than a microprocessor but only in high volume applications.

There are three main types of ASICs: the gate array, the standard cell, and the full custom[3]. The **gate array** is a chip made up of basic cells, each containing a number of unconnected transistors and resistors. Using a set of logic functions, known as a **cell library**, provided by the designer, the interconnection masks are generated. A designer can choose from a standard cell ASIC, consisting of a pre-characterized collections of gates, or a full custom ASIC, in which an individual description of each transistor occurs in building the circuit and the designer can specify each gate for optimum performance. Most manufacturers maintain cell libraries to simplify the design process of ASICs. A full custom implementation performs much faster than a standard cell; however, it costs significantly more. Also, recently the structured ASIC design was introduced, and is bridging the gap between field programmable gate arrays and the standard cell ASIC designs[2].

An **Application Specific Standard Product (ASSP)** is a semiconductor device, integrated-circuit product that is dedicated to a specific application market and sold to more than one user (standard). The ASSP is marketed to multiple customers, just as a general-purpose product is, but to a smaller number of customers since it is for a specific application. Like an ASIC, the ASSP is for a special application, but it is sold to any number of companies. (An ASIC is designed and built-to-order for a specific company[4].)

An ASSP generally offers the same performance characteristics and has the same die size as an ASIC. According to a Dataquest study, 17% of all semiconductor products sold in 1999 were ASSPs; 83% were general-purpose, showing a trend toward more application-specific products[4].

# The Product

## Application and End Uses

It is not uncommon to find logic circuits in a wide range of today's digital electronic equipment in industries such as telecommunications, computing, industrial, automotive industry, medical technology, internet applications, and consumer and military/avionics systems. While logic components can be found everywhere, the following are examples of specific products that may use logic circuits in various industries[10]:

- *Telecommunications* – Telephone systems, switching stations, electronic telephone books, mobile telephone/internet systems.
- *Computing* – Mobile computers, laptops, notebooks, mobile data input devices, etc.
- *Industrial* – Industry control technology, machine controllers, programmable logic controllers (PLCs).
- *Automotive Industry* – Navigation systems, passenger entertainment, onboard internet terminals.
- *Medical Technology* – Respiratory equipment, heart monitors, blood analysis equipment, brain activity analyzers, X-ray equipment, computer-aided graphs, data loggers.
- *Internet Applications* – Web terminals, set-top boxes, embedded web servers, digital TV with integrated internet browsers.
- *Consumer* – Point-of-sale, point-of-interest checkout systems, ATMs, vending machines, information terminals.
- *Military/Avionics* – Aircraft navigation, flight calculators for unmanned flight equipment, communications servers, rugged military applications.
- *Other* – Digital cameras and camcorders, video games and consoles, audio players, DVD players and recorders, televisions.

## Product Features and User Benefits

Very crude and rudimentary logic components hit the market in the 1950s. These components were bulky, slow, fragile, and required a large amount of power. However, in the past 50 years, logic components have changed dramatically.

Practically everything in today's logic component differs from the first components. For example today's logic components are smaller, faster, cheaper, more durable, more dependable, use far less power, can withstand higher temperatures, and are more resistant to interference. The miniaturization of current logic components is an important factor in today's market. Take cell phones for example; not only are current cell phones significantly smaller than those of ten years ago, but they are also more complex.

Modern logic components are far more efficient than their predecessors. With less power consumption, electronic products not only have longer battery life but also use smaller batteries, which results in smaller, lighter products. Because of new advances in manufacturing, cost-per-unit has decreased significantly over the past three decades. Logic components are also cheap because they are a commodity product and are mass-produced. Another improvement involves temperature, which is a very important factor in the performance of the chips. For example, the more power a chip uses, the hotter it operates and the slower and potentially less stable it operates. Temperature operating capacity continues to increase due to new heat resistant materials, such as ceramic. As logic components increase in speed, product complexity will continue to soar. Cell phones ten years ago were used for one thing—mobile calls. But today's cell phone applications have dramatically increased; for example, cell phones can be used for email, games, taking photos, web surfing, and Bluetooth data transfer.

Logic chips have increased in durability and dependability, which allows for a more stable system. Durability and dependability are very important in electronic devices, such as life support systems, and those used in the aviation and aerospace industries.

**Figure 1**: Product Features and User Benefits of Programmable Logic[1]

| Product Features | | User Benefits |
|---|---|---|
| Decreased Power Usage | ⇨ | Less power consumption requires smaller power source or less battery power. This enables products to be smaller and more compact. |
| Lower Operating Temperature | ⇨ | More efficient with lower overall operating temperature. |
| Variety of Speed Sorts | ⇨ | Enables user to select the device that offers the ideal speed for the function – trading off speed for cost. |
| Miniaturization | ⇨ | Results in drastically smaller and more complex electronic products. |

## Product Trends

Digital logic is operating in an extremely dynamic environment, and electronic equipment is used in a multitude of applications. Over the past few years, we have seen the logic industry increase each year. There is no evidence that would lead us to believe the market is heading toward a market plateau or decline in the near future. In fact, based on a recent market analysis, four important trends indicate a promising future for digital logic.

High performance is a key issue for today's market. The requirements have increased due to complicated graphics, communication interfaces for the internet, new operating systems, and other tools that are demanding increased performance and larger memory capacity. Faster, more connected, and more special features are the clues[5].

The consumer's demand for handheld and/or portable devices is growing, and this demand is driving the need for low-power devices. Some of the latest technology is geared toward an overall reduction of power consumption, while prolonging the operating time of battery-powered mobile systems, all with lightweight and smaller batteries. Also, low-power consumption decreases the board's operating temperature while increasing product reliability and the average elapse time that passes before failure, or mean time before failure (MTBF)[5].

Increasing performance requirements and smaller devices are making it trickier to reduce heat in embedded systems. Passive cooling, meaning not using a fan, is crucial for the reliability of the system. Unfortunately, it provokes pollution by attracting dust and other particles from the environment and it costs money. The advantages of passive cooling include a much longer MTBF time[5].

Having an attractive compact design can be compared to an organization method in which functions are combined into smaller packages A compact design will save board space, which is of great value, especially in the design of mobile systems[5].

Based on these trends, digital logic has many chances to enhance itself and increase the market. In the near future when these trends are met, new trends and expectations will arise and logic will once again prove a need for itself in the electronics market.

## Life Cycle

**Figure 2:** Life Cycle Curve for Programmable Logic Devices

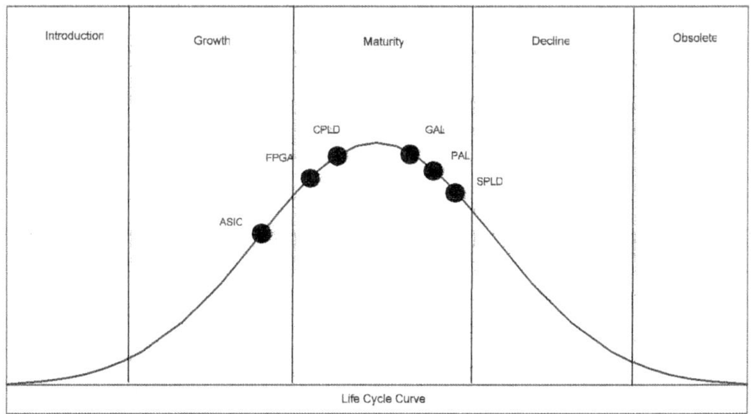

# The Market

## Market Trends

According to Gartner, the programmable logic market was approximately $3.1 billion in 2004. Within this market, there are two main segments: programmable logic devices ("PLDs") and field programmable gate arrays ("FPGAs"), each representing a distinct silicon architectural approach[10].

**Figure 3:** ASIC, ASSP, and FPGA/PLD Total Available Market, 2002-2008 ($M)[9]

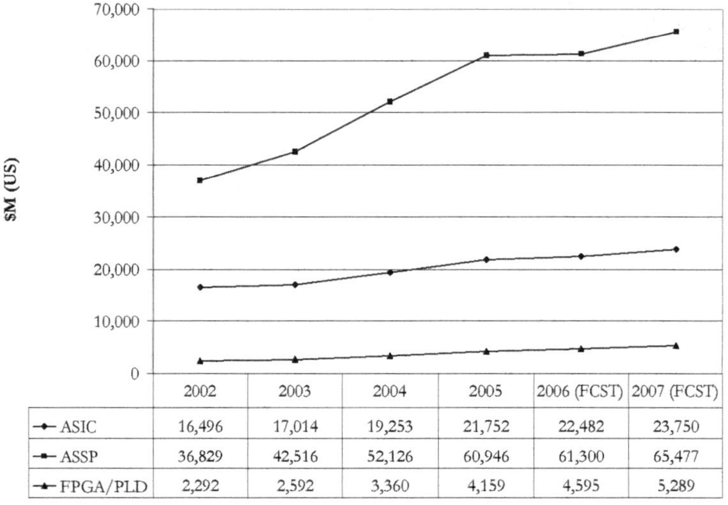


| | 2002 | 2003 | 2004 | 2005 | 2006 (FCST) | 2007 (FCST) |
|---|---|---|---|---|---|---|
| ASIC | 16,496 | 17,014 | 19,253 | 21,752 | 22,482 | 23,750 |
| ASSP | 36,829 | 42,516 | 52,126 | 60,946 | 61,300 | 65,477 |
| FPGA/PLD | 2,292 | 2,592 | 3,360 | 4,159 | 4,595 | 5,289 |

The market trend for programmable logic devices is forecasted to steadily increase as of a result of strong demands in industry, such as improvements on and more compact cellular phones and computers.

## Market Segmentation

Digital logic has shown a strong presence and resolute focus on fast-growing, high-margin segments such as telecommunications, medical technology, internet applications, and the automotive industry. Digital logic intends to achieve global leadership in many, if not all of these areas. Currently holding a firm grip on the European market, digital logic is systematically expanding its position in the U.S., and also serving the world's remaining regions through a network of distributors[3].

**Figure 4:** Examples of Digital Logic Applications in Each Market Segmentation[3]

| | |
|---|---|
| **Audio** | **Military** |
| AV Receivers | Avionics |
| Computing | Data Communications |
| Digital Radio | Electronic Countermeasures |
| Home Audio | Imaging Systems |
| Flat Panel | Munitions |
| Internet Audio | Navigation |
| Wireless | Radar/Sonar |
| **Automotive** | **Security** |
| Body Systems | Biometrics |
| Chassis Systems | Intelligent Sensing |
| Driver Information/Telematics | **Telecom** |
| Powertrain | HF Radios |
| Safety Systems | Infrastructure Equipment |
| Security Systems | Navigation Systems |
| **Broadband** | Telecom Accessories |
| 802.11 Wireless LAN | Wired Terminals |
| Cable Solutions | **Video and Imaging** |
| DSL Solutions | DLP™ Technology |
| VoIP Solutions | Digital Still Cameras |
| **Control** | Digital TV |
| Digital Power Supply | Digital Video Recorders |
| Embedded Sensing & | IP Video Phone |
| Measurement | Portable Media Devices |
| Industrial Drives | Set-Top Box |
| Motor Types | Streaming Media |
| Ultra-low Measurement | Surveillance IP Cameras |
| White Goods | Video Conferencing |
| **Medical** | Video Infrastructure |
| AED | **Wireless** |
| Biophysical Monitoring | Handsets and PDAs |
| Digital Hearing Aids | Infrastructure Equipment |
| Medical Imaging | RF Identification |
| Personal Medical Devices | Ultra Wideband |
| **Optical Networking** | |
| Optical Layer Applications | |
| Physical Layer Applications | |

# The Players

## Key Players

**Figure 5:** Top Seven Worldwide FPGA/PLD Market Share Ranking and Revenue, 2004 ($M)[9]

| | Supplier | Website | Distributors | Revenue $M (US) | Market Share % |
|---|---|---|---|---|---|
| 1 | Xilinx | www.xilinx.com | Avnet; NuHorizons | $1,586 | 51% |
| 2 | Altera | www.altera.com | Arrow; Digi-Key; Future | $997 | 32% |
| 3 | Lattice Semiconductor | www.latticesemic.com | Arrow; Avnet; Mouser | $226 | 7% |
| 4 | Actel | www.actel.com | Avnet | $166 | 7% |
| 5 | Cypress Semiconductor | www.cypress.com | Arrow; Avnet; Digi-Key; Future; | $50 | 2% |
| 6 | Quick Logic | www.quicklogic.com | Avnet; Future | $42 | 1% |
| 7 | Atmel | www.atmel.com | All American; Arrow; Avnet; Jan Devices | $15 | 0.2% |
| | **Total** | | | **$3,082** | |

**Figure 6:** Top 10 Worldwide ASIC Market Share Ranking and Revenue, 2004 (Millions of Dollars)[9]

| | Supplier | Website | Distributors | Revenue $M (US) | Market Share % |
|---|---|---|---|---|---|
| 1 | Texas Instruments (TI) | www.ti.com | Arrow Avnet; Digi-Key; Newark InOne; Rochester | $3,267 | 17% |
| 2 | IBC Microelectronics | www.ibm.com/chips | Arrow; Avnet | $2,125 | 11% |
| 3 | ST Microelectronics | www.st.com | Arrow; Avnet; Digi-Key; Future; Mouser; NuHorizons | $2,038 | 11% |
| 4 | NEC Electronics | www.nec.com | Arrow; Edge; Future Active Components; Future; Jaco; Symmetry | $1,674 | 9% |
| 5 | Fujitsu | www.fujitsu.com | Arrow; Bell Microproducts | $1,374 | 7% |

| | Supplier | Website | Distributors | Revenue $M (US) | Market Share % |
|---|---|---|---|---|---|
| 6 | Toshiba | www.toshiba.com | Arrow; Avnet; NuHorizons; TAEC | $1,027 | 5% |
| 7 | Agere | www.agere.com | Almac; Anterm; Arrow; Avnet; Echips | $967 | 5% |
| 8 | LSI Logic | www.lsilogic.com | Arrow; Avnet; Ingram Micro; MA Labs; MicroLand; Tech Data | $741 | 4% |
| 9 | Renesas Technology | www.renesas.com | All American; Arrow; Avnet; Digi-Key; NuHorizons | $613 | 3% |
| 10 | Samsung | www.samsung.com | Arrow; Edge; Future Active Components; Future; Jaco; Symmetry | $588 | 3% |
| | | Total | | $19,379 | |

## References


1 Altera. (2004). Retrieved Oct. 14, 2004 from <http://www.altera.com>.

2 Application-specific integrated circuit. (2005). Retrieved Jul. 06, 2005, from Wikipedia website: <http://en.wikipedia.org/wiki/Application-specific_integrated_circuit>.

3 ASIC. (2005). Retrieved Apr 19, 2005, from Texas Instruments website: <http://focus.ti.com/general/docs/asic/asichomepage.tsp?DCMP=TIHome Tracking&HQS=Other+OT+home_p_allasic>.

4 ASSP. (2003). Retrieved Jul. 06, 2005, from What Is It website: <http://whatis.techtarget.com/definition/0,289893,sid9_gci555538,00.html>.

5 Digital Logic. (n.d.). Retrieved Jul. 06, 2005 from <http://www.digitallogic.ch>.

6 Fairchild Semiconductor. (2004). Retrieved Oct. 1, 2004 from <http://www.fairchildsemi.com/>.

7 Field-Programmable Gate Array. (2005). Retrieved Jul. 06, 2005, from Wikipedia website: <http://en.wikipedia.org/wiki/FPGA.>.

8 Floyd L., T. (2002). *Digital Fundamentals*. New Jersey: Prentice Hall

9 Lewis, B. (2005). Forecast: ASIC/ASSP, FPGA/PLD and SLI/SOC Applications, Worldwide, 2002-2010, 1Q05 Update (G00126883), Retrieved Jul 06, 2005, from Gartner database.

10 Programmable Logic Device. (2005). Retrieved Apr 14, 2005 from Answers website: <http://www.answers.com/topic/programmable-logic-device?method=6>

11 Programmable Logic Device. (2005). Retrieved Jul 06, 2005, from Wikipedia website: <http://en.wikipedia.org/wiki/Programmable_logic_device>.

12 Programmable Logic Market Background. (2005). Retrieved Jul. 06, 2005, from Lattice Semiconductor website: <http://www.latticesemi.com/investors/profile/background.cfm?>.

# 14

# Memory

## Volatile & Nonvolatile

**Memory**, in electronics, is used to internally store data. Computer memory is only capable of remembering sequences of zeros and ones, but by utilizing the binary number system, it is possible to produce arbitrary rational numbers and through formatting, all representations of pictures, sounds, and animations can be made[13]. There are two basic types of memory: volatile and non-volatile. Volatile memory is a memory whose contents are permanently lost when power is removed; RAM (random access memory) is an example. Non-volatile memory is just the opposite—when power is removed, the contents are stored and can be recovered; examples are ROM (read-only memory) and Flash memory. Figure 1 shows the hierarchy of MOS (metal-oxide semiconductor) memory.

**Figure 1**: The Hierarchy of MOS Memory[1]

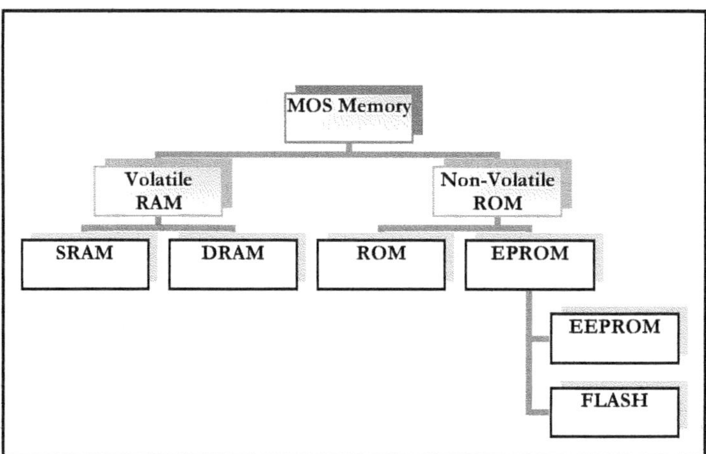


**Volatile memory** is a type of memory that loses its contents when the power is turned off. All forms of RAM, except for CMOS RAM, are considered non-volatile. RAM stands for "random access memory" and this means that any byte of memory can be accessed without accessing the preceding bytes[13]. RAM is considered the most common type of memory found in computers and other electronic devices. RAM comes in two different types: DRAM, dynamic random access memory, and SRAM, static random access memory. The two types differ in the technology they use to store data.

**DRAM**, dynamic random access memory, is a type of random access memory that stores each bit of data in a separate capacitor. "Dynamic" indicates that to remember data, every bit in the memory chip mush be refreshed within a certain time period or it will lose its contents[8]. DRAM must refresh frequently due to a leak of charge causing the capacitor to decay. This amount of electric charge also determines whether the bit is considered 1 or 0. "Random access" indicates that locations in the memory can be written to or read from in any order, regardless of the memory location that was accessed in previous operations. DRAMs use address multiplexing, a technique that splits the address in half and feeds each half in turn to the chip on the same set of pins. Therefore, DRAM needs half as many addresses, enabling it to be packaged using smaller and cheaper options[12].

The alternative to DRAM is **SRAM**, static random access memory. "Static" indicates the memory does not need to be refreshed as frequently as DRAM does. However, once the power source has been removed, SRAM looses it memory[13]. Unlike multiplexed DRAMs, SRAMs have a perfectly symmetrically circuit structure which allows the value of a memory location to be read much faster then DRAM. The cycle length of SRAM is much shorter than that of DRAM because it does not need to pause between accesses[10]. SRAM is more involved technology and is more expensive to produce, therefore it has a higher cost. Yet it is a better type of random access memory than DRAM[13].

**Non-volatile memory** is the second category of memory. Memory that is considered non-volatile will retain its contents when power is removed. Most forms of ROM, or read-only memory, are non-volatile. ROM means that the data is prerecorded to the memory chip; once this happens, the information cannot be removed and can only be read[13]. The three most basic forms of ROM include EPROM, EEPROM and Flash memory. EEPROM and Flash are both considered non-volatile; however, since their technology is so much more advanced compared to DRAM, SRAM and EPROM, they are often considered hybrids of RAM and ROM[1].

**EPROM** is a special type of memory that retains its contents until it is exposed to ultraviolet light[13]. The ultraviolet light clears its contents, making it possible to reprogram the memory. To write to and erase an EPROM, you need a special device called a PROM programmer or PROM burner. An EPROM differs from a PROM (Programmable Read-Only Memory) in that a PROM can be written to only once and cannot be erased. EPROMs are used widely in personal computers because they enable the manufacturer to change the contents of the PROM before the computer is actually shipped. This means that bugs can be removed and new versions installed shortly before delivery[9].

Electrically erasable programmable read-only memory, **EEPROM**, is a form of read-only memory than can be electrically erased and then reprogrammed while in place. It is referred to as "ee-ee-prom" or "E-square prom." EEPROMs are very similar to EPROM and Flash; however, they can be erased and programmed not only externally but also within a computer[9]. However, EEPROMs cannot be written to as a whole but must be written byte by byte. EEPROMs further differ from EPROMs because they can be erased and reprogrammed multiple times—typically 100,000 to 1,000,000[13]. An EEPROM's function is to store programs and data that are never, or only infrequently, modified. EEPROMs are also considered a non-volatile memory chip, which means that when power is removed, the data is still kept within the chip. There are two types of EEPROM:
**Serial –** These EEPROMs are used to store personal preference data and configuration, or setup, data. They are the most flexible type of non-volatile memory utilized[14].
**Parallel –** This type of EEPROM enables the user of applications to update the stored data, byte-by-byte or full sector while providing reliability benefits of high programming endurance, data retention, and faster read timing than serial EEPROMs[14].

**Flash memory** is a type of constantly powered nonvolatile memory that can be erased and reprogrammed in units of memory called blocks. Flash memory stores information on a silicon chip in a way that does not need power to maintain the information in the chip. This means if you turn the power off to the chip, the information is retained without consuming any power[7]. Flash memory gets its name because the microchip is organized so that a section of memory cells are erased in a single action of "flash." There are several types of Flash memory available on the market and they consist of:

**NOR**- These cells represent the earliest Flash design. NOR Flash memory has a large sector size and cell size, high-speed rewrite and high-speed random data access. Storage capacity for NOR chips is currently 64 Kb to 8 Mb. It is ideal for code storage applications like PC Bios and cellular phones[7].

**NAND**- Serial access devices that are characterized by high storage capacity, single voltage operation, small block size, small cell size, and high durability. Storage capacity is currently in the range of 500 Kb to 8 Mb. This type of memory is more suitable for applications that do not require repetitive random access like MP3 players and digital cameras[7].

**DINOR** (Divided Bit-Line NOR)- The chips have random data access, single voltage supply fast block erase, large block size, multi-sector erase capabilities, and low power consumption. Storage capacity is currently in the range form 1 MB to 2 MB[3].

**AND**- Structures are serial access devices combining the best of the NOR and NAND devices. It has fast erase, single voltage supply operation, very high capacity, low power consumption, and small block size[3].

# The Product

## Applications and End Uses

DRAM is found in many different segments—computing, infrastructure, and entertainment. Desktop PCs, notebook PCs, workstations, graphics, thin clients, and peripherals, such as scanners and printers, are all areas of computing that use DRAM. In infrastructure, DRAM is found in servers and networking products including routers, switches, and storage devices. Finally, DRAM can be found in entertainment devices such as game consoles, set-top boxes, camcorders, DVD players, DVD-ROMs, CD players, digital televisions, digital still cameras, digital projectors, and ebooks[1].

SRAM is used in many different applications, including PC computers, disk drives, workstations, and servers. In the consumer industry, it is found in game consoles, digital cameras, and internet appliances. In communications, SRAM is used for cell phones and base stations and also in networking equipment such as switches, routers, and data line cards. SRAM can also be seen in residential data gateway, which includes set-top boxes, DSL modems, and cable modems, and lastly in disk drive data storage[1].

Non-volatile memories, EPROM, EEPROM and Flash, are used in most of the segments of the electronics component market. They are found in everyday household items, such as cordless phones, answering machines, and base stations. They are also found in most media playing devices. For example, CD players, and VHS and DVD machines. Video games use memory devices in their DVD high density functions. They also are found in the everyday workings of corporations, such as networking and printer solutions. Currently, memory is popularly being used in digital cameras, MP3 players, portable digital assistants, internet, video game consoles, digital cellular phones, LAN switches, PC cards for notebook computers, answering machines, and digital set-up boxes[1].

## Product Features and User Benefits

**Figure 2:** Memory Features and User Benefits[1]

| Product Features | | User Benefits | Applicable Memory Types |
|---|---|---|---|
| Write able | ⇨ | Can be changed without having to buy a new chip. | DRAM SRAM EPROM EEPROM Flash |
| Unlimited Erase Cycles | ⇨ | Can be written to an infinite number of times. | DRAM SRAM Flash |
| Low Power Consumption | ⇨ | Doesn't require its own battery or very powerful computer. | DRAM SRAM |
| Fast Read Time | ⇨ | Doesn't take a lot of time to obtain information on chip or to change information on chip. | SRAM EPROM EEPROM Flash |
| Fast Write Time | ⇨ | Doesn't take long to change information on chip. | SRAM Flash |
| Temperature Withstanding | ⇨ | Can withstand most industrial and commercial temperatures. | SRAM Flash |
| Memory Expansion | ⇨ | Simple process to add memory inputs. | SRAM |
| Non-Volatile | ⇨ | Stores information after power is lost. | EPROM EEPROM Flash |
| High Densities | ⇨ | Allow for use of applications such as gaming devices, VCRs, DVDs, printers, and fax machines. | EPROM EEPROM Flash |
| Reprogrammable System | ⇨ | Doesn't need to be removed from the system to be programmed. | DRAM SRAM EEPROM Flash |
| Bit by Bit Erasure | ⇨ | Doesn't have to erase the whole chip, can erase bit by bit. | DRAM EEPROM Flash |

## Product Trends

There are most likely two directions for the future of memory products: holographic memory and molecular memory. **Holographic memory** is a type of optical recording that can have high record density and also maximum data access speed. The data is recorded and coded in one big block at one access. The advantages are that holographic memory will have very high data rates and high-speed random access, as well as almost no mechanical components, which will give it a longer life, lower power consumption, and less probability of failures[11]. The other, **molecular memory**, is a memory system that will use the digital bits of a molecule called bacteriorhodopsin. It is a protein molecule that absorbs light, therefore lasers are used to erase and change the information. An advantage of molecular memory includes the fact that it is protein based, which comes in large volumes at low prices. Also, the memory system is able to operate in a more extensive range of temperatures; and finally, data is stored constantly even if power is removed. This memory product also may be more convenient than discs or cartridges because the bricks are small in size but contain large amounts of data and can be placed into archive for data storage[11].

## Life Cycle

The timeline below denotes the major landmarks in the memory and semiconductor industry. The Life Cycle Curve illustrates the positioning of each of the memory types in their product life.

**Figure 3:** Timeline of Electronic Memory Products

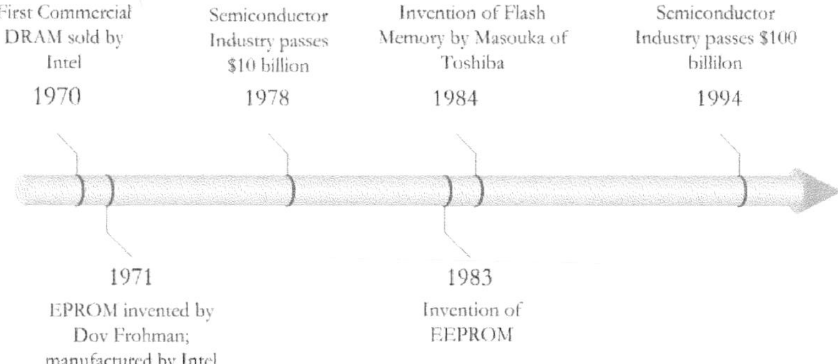

**Figure 4:** Life Cycle Curve for Memory Products

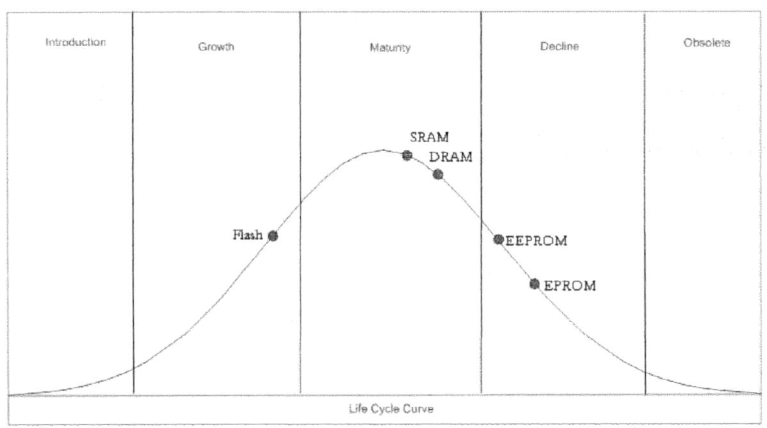


# The Market

## Market Trends

**Figure 5:** Total Revenues and Future Trends for Memory[4, 6]

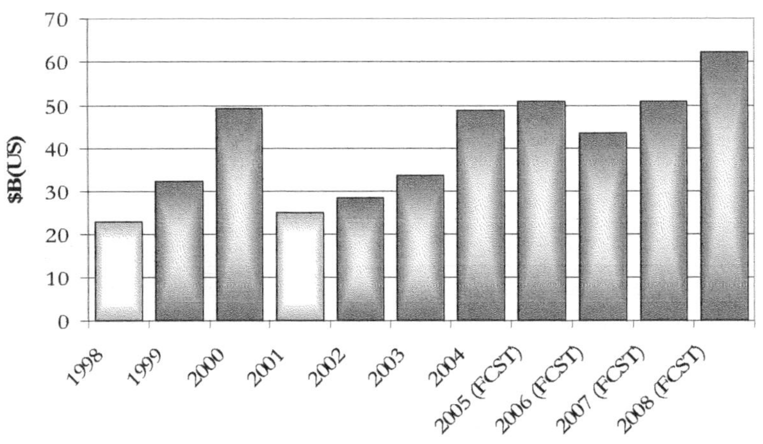


| 1998 | 1999 | 2000 | 2001 | 2002 | 2003 | 2004 |
|---|---|---|---|---|---|---|
| $23 | $32.30 | $49.20 | $24.90 | $28.40 | $33.70 | $48.70 |

| 2005 (FCST) | 2006 (FCST) | 2007 (FCST) | 2008 (FCST) |
|---|---|---|---|
| $50.80 | $43.50 | $50.80 | $62.30 |

It has been suggested that the downturn seen in the memory market during 1998, 2001, and what is forecasted for 2006 can be directly related to the cyclical nature of memory. Memory pricing has a tendency to fluctuate dramatically and this, combined with demand supply dynamics, tends to drive interesting market trends. Economists suggest that in these years supply exceeds demand[4].

## Market Segmentation

**Figure 6:** Memory Revenue Segmentation by Technology, 2004[5]

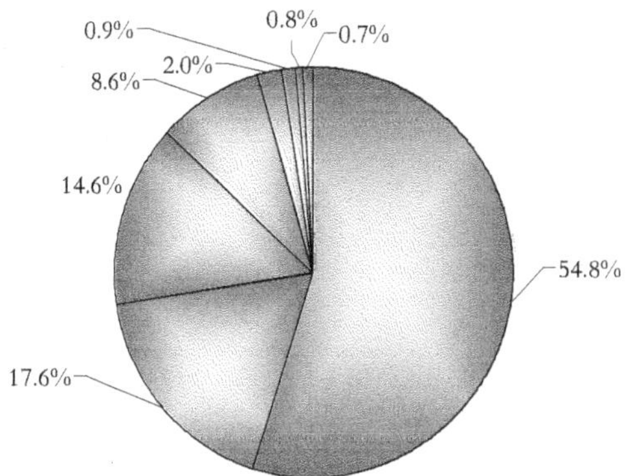


| Memory Technology | Revenue |
|---|---|
| DRAM | 54.8% |
| NOR Flash | 17.6% |
| NAND Flash | 14.6% |
| SRAM | 8.6% |
| EEPROM | 2.0% |
| Other Memory | 0.9% |
| EPROM | 0.8% |
| Mask ROM | 0.7% |

**Figure 7:** Flash Memory Market Segmentation[2]

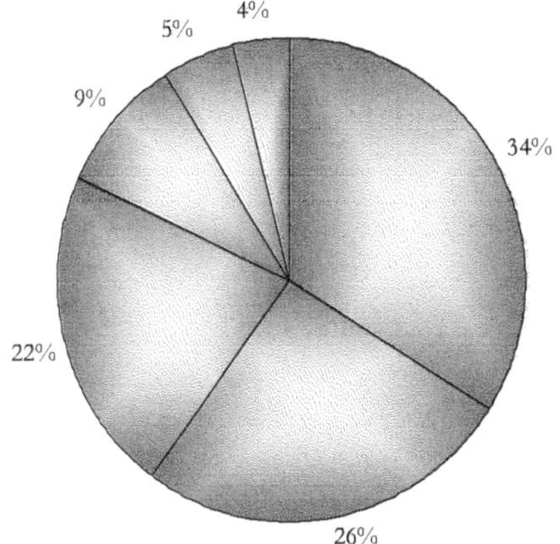


| Industry | Market Share % |
|---|---|
| Telecom | 34% |
| Consumer | 26% |
| Computers | 22% |
| Transportation | 9% |
| Industrial | 5% |
| Military | 4% |

# The Players

## Key Players

**Figure 8:** Top Manufacturers of Memory by Revenue, 2004[5]

| | Supplier | Website | Distributors | Revenue $B (US) | Market Share % |
|---|---|---|---|---|---|
| 1 | Samsung | www.samsung.com | All American; Arrow; Empower Labs; Jaco; MyMemoryStore; Reptron | $13.49 | 28.1% |
| 2 | Hynix Semiconductor | www.hynix.com | Hynix Semiconductor America | $4.65 | 9.7% |
| 3 | Micron Technology | www.micron.com | Arrow; Avnet; Edge; Semi Dice | $4.37 | 9.1% |
| 4 | Infineon Technologies | www.infineon.com | Arrow; Avnet; Digi-Key; Future; Richardson | $3.62 | 7.5% |
| 5 | Toshiba | www.toshiba.com | Arrow; Avnet; NuHorizons; TAEC | $2.44 | 5.1% |
| 6 | Spansion | www.spansion.com | Arrow; Avnet; Future | $2.34 | 4.9% |

| | Supplier | Website | Distributors | Revenue $B (US) | Market Share % |
|---|---|---|---|---|---|
| 7 | Intel | www.intel.com | Arrow; Avnet; Digi-Key; Rochester | $2.29 | 4.8% |
| 8 | ST Microelectronics | www.st.com | Arrow; Avnet; Digi-Key; Future; Mouser; NuHorizons | $1.65 | 3.4% |
| 9 | Elpida Memory | www.elpida.com | Symmetry Electronics Corp. | $1.57 | 3.3% |
| 10 | Nanya Technology | www.nanya.com | Avnet | $1.19 | 2.5% |
| **Total** | | | | **$48.03** | |

## References


1. Barr, M. (2001). Introduction to Memory Types. Retrieved Jun. 28, 2005, from Embedded Systems Dictionary website: <http://www.netrino.com/Publications/Glossary/MemoryTypes.html>.

2. Carbone, J. (2003). Expect a Price War for NAND Flash as Demand Grows. *Purchasing*, 132. Retrieved Jun 28, 2005, from TableBase database.

3. Daniel. (1999). Flash Memory Defined. Retrieved Jun. 28, 2005, from CST Publications website: <http://www.ddrtester.com/page/news/showpubnews.asp?title=Flash+Memory+Defined&num=10>.

4. Dram Market Faces Structural Weakness. (2001). Retrieved Jun. 28, 2005, from Thin Film Manufacturing website: <http://www.thinfilmmfg.com/subscribers/Subscriber01/dram20Nov01.htm>.

5. Gordon, R. (2005). Market Share: Memory, Final Rankings, Worldwide, 2004 (G00127414). Retrieved Jun 28, 2005, from Gartner database.

6. Gordon, R., & Norwood, A. (2004). Forecast: Memory, Worldwide, 1998-2008, 4Q04 Update (G00125382), Retrieved Jun 28, 2005, from Gartner database.

7. How Flash Memory Works. (2005). Retrieved Jun. 28, 2005, from Byte website: <http://www.netrino.com/Publications/Glossary/MemoryTypes.html>.

8. Kozierok, C. M. (2001). Dynamic RAM (DRAM). Retrieved Jun. 28, 2005, from Dynamic RAM (DRAM) website: <http://www.pcguide.com/ref/ram/typesDRAM-c.html>.

9. Kozierok, C. M. (2001). Static RAM (SRAM). Retrieved Jun. 28, 2005, from Dynamic RAM (DRAM) website: <http://www.pcguide.com/ref/ram/typesSRAM-c.html>.

10. Kozierok, C. M. (2001). Read-Only Memory (ROM). Retrieved Jun. 28, 2005, from Dynamic RAM (DRAM) website: <http://www.pcguide.com/ref/ram/typesROM-c.html>.

11. Len, M. (n.d.). Memory of the future: two directions. Retrieved Jun. 28, 2005, from Digit-Life website: http://www.digit-life.com/articles/memorytwodirections/.

12. Memory, A.C. (2005). How Memory Works. Retrieved Jun. 28, 2005, from Transtronics Web site: <http://www.xtronics.com/memory/how_memory-works.htm>.

13. Memory (n.d.). Retrieved Jun. 28, 2005, from Webopedia website: <http://www.webopedia.com/TERM/m/memory.html>.

14. Tyson, T. (1993). Serial EEPROM Solutions v. Parallel Solutions. Retrieved Jun. 28, 2005, from Microchip website: <http://www.eetasia.com/ARTICLES/2000JUN/2000JUN07_MEM_AN4.PDF>.

# 15 Optoelectronics

**Optoelectronics** deals with light (optical energy) that is converted into electricity (electrical energy) and with electrical energy that is converted into optical energy. Optoelectronics can also be referred to as **photonics**. Items that convert optical energy or photons into electrical energy or electrons are passive optical components, such as optical fibers, waveguides, amplifiers, attenuators, and splitters. Light emitting diodes (LED) and lasers are light sources and optoelectronic components that convert electricity into light[22]. Typically, optoelectronic devices are divided into four categories, as shown in Figure 1.

**Figure 1:** Four Categories of Optoelectronic Devices[22]

| Device | Definition | Examples |
|---|---|---|
| Light Sensing Devices | Incoming light produces an electrical output. | Photocells<br>Photodiodes<br>Phototransistors |
| Light Emitting Devices | An electrical input produces a light output. | Light-emitting diodes (LED)<br>Laser diodes |
| Power Conversion Devices | A light input is converted into electrical power. | Solar cell |
| Electro-Optical Systems | A device that electrically isolates two related circuits, usually a transmitter and a receiver, and transfers the power between them. | Optocouplers (Opto isolators,<br>Photon coupled isolators,<br>Photocouplers,<br>Photo coupled pairs,<br>optically coupled pairs) |

The phenomenon called photoconductivity is the basis for the photoconductive cell, or photocell, the simplest of all optoelectronic devices. A **photocell** is a type of resister that allows current to flow freely when light is hitting it. However, when no light is present, the resistance in the photocell increases, limiting free current flow. Photocells can be used to detect small light fluctuations, such as between two light bulbs, or very large light fluctuations, even between direct sunlight and total darkness[17].

A **photodiode** is a two-electrode, positive-negative (PN) junction that is formed in a semiconductor. As light hits the junction, the current reverses direction[2]. A photodiode is the reverse of an LED. Photodiodes collect the signal carried by a modulated light beam and then in response produce a modulated current[11]. Two common uses of the photodiode are the detection of optical power and the conversion of optical power to electrical power. Photodiodes are used to measure responses in x-ray, ultra violet (UV), visible, and infra-red (IR) spectrums. X-ray photodiodes are used to detect x-ray, gamma ray, and beta radiation wavelengths. The UV photo diode detects both the UV and the blue spectral region, or **wavelength range**. UV and IR photodiodes require enhanced fabrication processes. Photodiodes are chosen based on performance specifications, including sensitivity, rise time, quantum efficiency, and operating temperature[2].

A **phototransistor** is an optoelectronic device that is similar to the transistor and the photodiode; it is controlled by light hitting the PN junction of the device. The difference from a photodiode is that a phototransistor offers built-in gain and also a greater sensitivity to light[12].

The **Light Emitting Diode (LED)**[1] is another PN junction device that radiates light when located in the forward direction. Most often they emit light in the near infrared and visible ranges; however there are LEDs that operate in the UV range as well. LEDs do not require much power and they generate very little heat, making them a much better and more reliable source of light compared to incandescent lamps and neon lamps. When chosen by performance, it is important to look at an LED's peak wavelength, viewing angle, optical power output, luminous intensity, forward current, and also the forward voltage. LED types include[1]:

- Infrared
- Red
- Orange
- Yellow
- Green
- Blue
- White
- Ultraviolet

A **laser diode** is a laser with a PN junction in a semiconductor that is used to create analog signals or digital pulses for transmission through optical fibers. LEDs are also used for this purpose; however, the difference is a laser diode will generate a smaller beam that is easier to couple with the core of single mode fibers. A laser diode can project light from both its edge and its surface. When the light is emitted from its surface, it creates a circular beam that couples easily with the round core of single mode fibers[8].

A **solar cell**, also called the **photovoltaic cell**, is another optoelectronic device that has a large-area PN junction in a semiconductor. Using the photovoltaic effect, in the presence of sunlight the solar cell is able to create usable electrical energy[23].

**Optocouplers** are optoelectronic devices that are used to transfer an electrical signal between two circuits while isolating the circuits from each other[14]. Usually, an optocoupler has an infrared LED section that can be either AC or DC at the input and a silicon photodetector at the output. The photodetector at the output can be a photocell, photodiode, or a phototransistor[14]. When an electrical signal is applied to the LED it lights up, the photodetector then activates and finally the output generates a corresponding electrical signal[13].

# The Product

## Application and End Uses

Photocells are often used as light meters for photographic purposes since the maximum spectral response of a photocell is in the visible region of the spectrum. Photocells are also used for applications such as turning on lights. However, since they have a very slow response time, they are not used for high-speed applications like optical communications[22].

The applications of photodiodes are similar to that of photocells. They can be found in camera light meters, clock radios if they dim the display when it's dark, and also streetlights. Many times the receivers for VCR and television remote controls use photodiodes. Photodiodes also are used for very accurate light intensity measurement because they generally have a more linear response than photocells; however, photodiodes are not used to measure extremely low light intensities[18].

Phototransistors are also used to detect light and are commonly seen in a camera to control the shutter. They are also found in computer and business equipment, such as floppy drives, printers, and copiers as well as industrial settings, such as light pens, security systems, and safety shields. In the consumer industry, phototransistors are found in coin counters and lottery card readers. They are also used in gaming devices, including joysticks, toys, audio/visual equipment and even in laser tag[24].

LEDs are used in many common applications similar to those of the photocell and the phototransistor. They are used as information indicators at airports and railway stations as well as the on/off lights on professional instruments and audio/video equipment. In addition, LEDs are used as movement sensors in mechanical and optical computer mice and trackballs. They are found in remote controls, flashlights, backlights for LCD displays, digital clocks, calculator displays, light up watches, and even the lights in an electric appliance. LEDs replaced the ordinary bulbs behind colored glass in traffic signals. In industries in which night vision is needed, red or yellow LEDs are used. They are found in photographic dark rooms, aircraft cockpits, submarine and ship bridges, astronomy observatories, and animal watching and military fields[9].

Laser diodes are used extensively in the telecommunications industry because they are easily modulated and coupled light sources for fiber optics. One common use is in measuring instruments like rangefinders and barcode readers. The visible laser in a laser pointer is typically a laser diode. Finally, red laser diodes are seen in CD players, CD-ROMs and DVD technology and blue laser diodes are found in new technologies like HD-DVD and Blu-Ray[3].

Solar cells are traditionally used in circumstances in which electrical power from the grid is not present, for example, in remote area power systems, orbiting satellites, handheld calculators, remote radiotelephones, and water pumping applications. Solar cells are beginning to be mounted on building roofs where they are connected to the electricity grid through an inverter and net metering arrangement[24].

Optocouplers are used in many control applications; specific examples include punch-card readers, parts counters, end-of-tape sensors on a printer or tape recorder, and as part of an interlock mechanism[4]. They also can replace and read relays because optocouplers give a faster switching speed with no contact bounce. In floating applications, optocouplers can replace pulse transformers. Optocouplers are also used to protect applications against noise voltages of up to several thousand volts[4].

## Product Features and User Benefits

**Figure 2:** Photocells[16]

| Product Features | | User Benefits |
|---|---|---|
| Responsive to Very Low and Very High Light Levels | ⇨ | The same photocell can be used in moonlight (very low levels) as well as sunlight (very high levels). |
| Wide Dynamic Range (Allows More Current to Flow) | ⇨ | Signal current from the detector can be viewed over a wide range by adjusting the applied voltage. |
| Low Noise Distortion | ⇨ | The amounts of fluctuations in voltages are low as well. Natural and deliberate interferences have little effect on the photocell's performance. |
| Maximum Operating Voltages of 50-400 Volts | ⇨ | Suitable for operation on 120/240 VAC. |

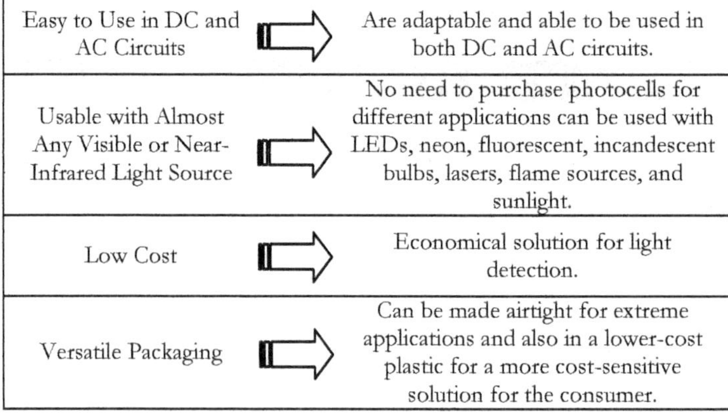

| Product Features | User Benefits |
|---|---|
| Easy to Use in DC and AC Circuits | Are adaptable and able to be used in both DC and AC circuits. |
| Usable with Almost Any Visible or Near-Infrared Light Source | No need to purchase photocells for different applications can be used with LEDs, neon, fluorescent, incandescent bulbs, lasers, flame sources, and sunlight. |
| Low Cost | Economical solution for light detection. |
| Versatile Packaging | Can be made airtight for extreme applications and also in a lower-cost plastic for a more cost-sensitive solution for the consumer. |

**Figure 3:** Photodiodes[19]

| Product Features | User Benefits |
|---|---|
| Low Cost | Economical solution for light-detection applications. |
| Excellent Linearity Output | Outputs follow predictable line in 7-9 decades of light intensity. |
| Fast Response Time | Will relay outputs in nanoseconds. |
| Low Noise | The amounts of fluctuations in voltages are low. Natural and deliberate interferences have little effect on the photodiode's performance. |
| Mechanically Rugged | Sturdy and strong in construction. |
| Compact | Small, does not take up a lot of space in the system. |
| Lightweight | Will not affect the weight of the system. |
| Usable with Almost Any Visible or Near-Infrared Light Sources | Solid-state laser diodes, neon, fluorescent, incandescent bulbs, lasers, flame sources, and sunlight. |

**Figure 4:** Phototransistors[20]

| Product Features | | User Benefits |
| --- | --- | --- |
| Low Cost | ⇨ | Economical solution for light-detection applications. |
| Moderately Fast Response Times | ⇨ | Not as fast as photodiode. |
| Versatile Packaging | ⇨ | Can be made airtight for extreme applications, in a lower-cost plastic for a more inexpensive solution, and also in chip form. |
| Usable with Almost Any Visible or Near-Infrared Light Sources | ⇨ | Solid-state laser diodes, neon, fluorescent, incandescent bulbs, lasers, flame sources, and sunlight. |
| Better Gains than Photodiode | ⇨ | Require less input voltage or power to acquire more output voltage or power. |

**Figure 5:** Light Emitting Diodes (LEDs)[10]

| Product Features | | User Benefits |
| --- | --- | --- |
| Emit Color Lights | ⇨ | Do not require the use of color filters. |
| Focusable Light | ⇨ | The shape of the LED package allows for the light to be focused; other light sources require an external reflector to collect and direct light. |
| Solid Packaging | ⇨ | Allows for LED to be protected as well as hard to fracture and very durable. |
| Most Have a 100,000 Hour Length Life Span | ⇨ | The life span of an LED is approximately twice as long as the best fluorescent bulb and 20 times longer than the best incandescent bulb. Replacement is less frequent than fluorescent and incandescent bulbs. |
| Energy Efficient | ⇨ | Most of the energy is created into light and not heat, and have efficiencies as high as ten times that of incandescent. |

**Figure 6:** Laser Diodes[3]

| Product Features | | User Benefits |
|---|---|---|
| Long Life | ⇨ | Replacement is less frequent than fluorescent and incandescent bulbs, and even LEDs. |
| Flexibility in Customization | ⇨ | Can be customized for most applications, good for long lengths of fiber optic systems. |
| High Efficiency | ⇨ | Most of the energy is created into light and not heat; more efficient than the LED. |
| Small Spectral Width | ⇨ | Can focus on smaller areas of the spectrum. |
| Small Emitting Area | ⇨ | Allows for the beam of light to be focused into an even smaller area. |

**Figure 7:** Solar Cells[21]

| Product Features | | User Benefits |
|---|---|---|
| Modular | ⇨ | Electrical power and energy can be engineered for virtually any application. |
| No Moving Parts | ⇨ | Makes the system reliable since there is nothing to generate friction and eventually wear out; also makes the cells generate no noise and have a long life. |
| Consume no Fuel | ⇨ | Low operating costs. |
| Versatile | ⇨ | Adding cells to a system will increase, or removing cells will decrease the amount of power generated. |
| Generate no Pollution | ⇨ | Good for the environment. |

**Figure 8:** Optocouplers[4]

| Product Features | | User Benefits |
|---|---|---|
| Economical | ⇨ | Low cost per output benefit, compared to relays and pulse transformers. |
| Solid-State Reliability | ⇨ | No moving parts that could generate friction or heat. No need to replace moving parts that wear out. |
| DC & Low Frequency Transmission | ⇨ | Ability to transmit signals through mediums that are not conducive to transmitting electrical signals. |
| High Voltage Isolation | ⇨ | Can be used to isolate circuits in high voltage applications. |
| Small Size Dual In-Line Package (DIP) | ⇨ | Comes in very small sizes and can be used in very small and large applications without taking up too much room in the system. |
| Low Power Operation | ⇨ | Does not require much power to operate the electro-optical system. |

## Product Trends

The trend in the optoelectronics market will be in enabled products, such as optical disk storage and flat panel displays. An **optoelectronic -enabled product** is any device that uses an optoelectronic component as part of its major function. That includes laptop computers, compact disk storage drives, and digital cameras. The major enabled products that are taking off include fiber-based telecommunications and data communications, optical disk drives, scanners and cameras, and finally the largest segment—computers and display equipment[12].

## Life Cycle

**Figure 9:** Life Cycle of Optoelectronic Components

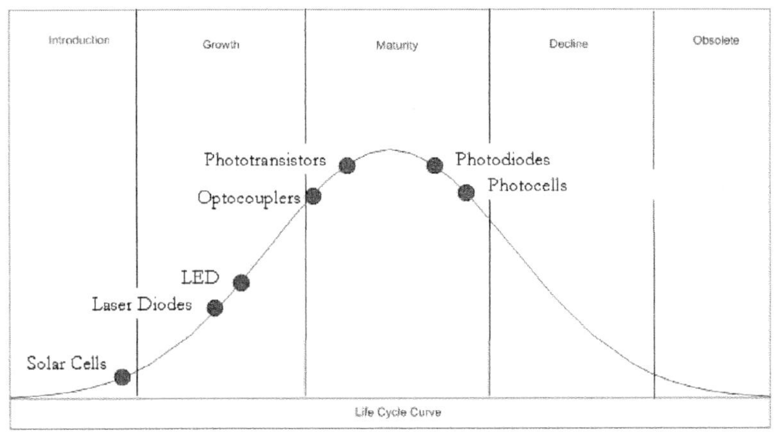


# The Market

## Market Trends

**Figure 10:** Revenues of Optoelectronics from 2002-2010[7]

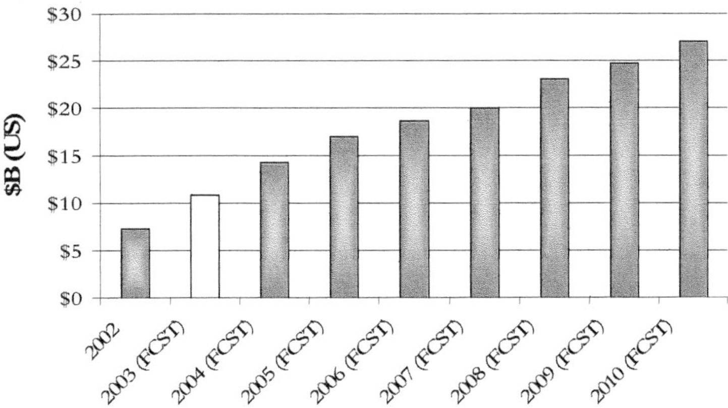

| | Optoelectronics | Total Semiconductor Revenue | Percent of Semiconductor Revenue (%) |
|---|---|---|---|
| 2002 | $7.3 | $155.6 | 4.69% |
| 2003 | $10.8 | $178.2 | 6.06% |
| 2004 | $14.3 | $220.9 | 6.47% |
| 2005 (FCST) | $17.0 | $232.9 | 7.3% |
| 2006 (FCST) | $18.6 | $248.1 | 7.5% |
| 2007 (FCST) | $20.0 | $260.8 | 7.6% |
| 2008 (FCST) | $23.1 | $294.1 | 7.7% |
| 2009 (FCST) | $24.7 | $299.1 | 8.4% |
| 2010 (FCST) | $27.0 | $318.4 | 8.5% |

## Market Segmentation

**Figure 11:** Market Segmentation by Application of Optoelectronic Semiconductors in 2000[12]

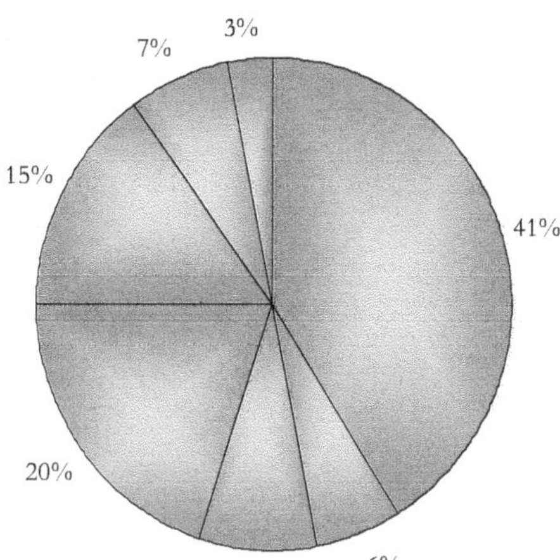


| Applications of Optoelectronic Semiconductors 2000 | |
|---|---|
| Telecom | 41% |
| Consumer | 20% |
| Computers | 15% |
| Industrial | 8% |
| Automotive | 7% |
| Military/Aero | 6% |
| Other | 3% |

**Figure 12:** Optoelectronics Enabled Products and their Market[12]

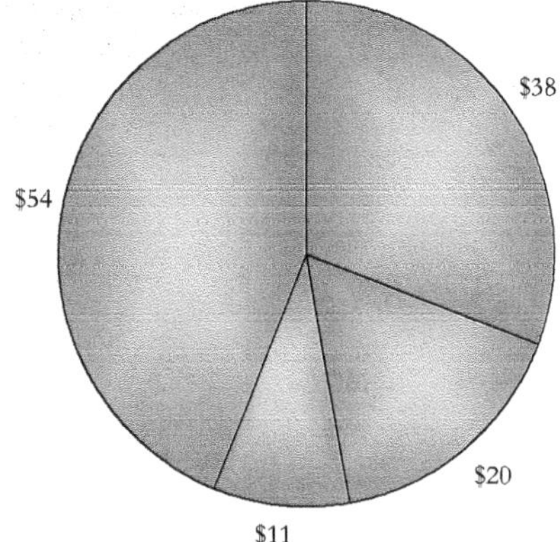


| Market Segmentation of Optoelectronics Enabled Products ($B US) | |
|---|---|
| Computers and Display Equipment | $54 |
| Fiber based Telecommunications & Datacommunications | $38 |
| Optical Disk Drives | $20 |
| Scanners and Cameras | $11 |

# The Players

## Key Players

**Figure 13:** 2003 Top 20 Key Players in Optoelectronics[7]

| | Supplier | Website | Distributors | Revenue $M (US) | Market Share % |
|---|---|---|---|---|---|
| 1 | Sharp | www.sharpusa.com | Active Light; Electrograph | $1,305 | 11.2% |
| 2 | Sony | www.sony.com | Bell Microproducts | $1,285 | 11% |
| 3 | Agilent | www.agilent.com | Arrow; Avnet; Digi-Key; Future; Newark InOne | $1,132 | 9.7% |
| 4 | Panasonic | www.panasonic.com | | $876 | 7.5% |
| 5 | Toshiba | www.toshiba.com | Arrow; Avnet; NuHorizons; TAEC | $652 | 5.6% |

| | Supplier | Website | Distributors | Revenue $M (US) | Market Share % |
|---|---|---|---|---|---|
| 6 | Nichia | www.nichia.com | | $447 | 3.8% |
| 7 | Sanyo | www.sanyo.com | | $444 | 3.8% |
| 8 | Citizen Electronics | www.c-e.co.jp/e | | $400 | 3.4% |
| 9 | Vishay | www.vishay.com | Allied; Arrow; Avnet; Future; Jaco; Passive Components; Projections Unlimited; PUI; Richardson; Semi Dice; Taitron; TTI | $398 | 3.4% |
| 10 | NEC Electronics | www.necelam.com | Avnet; Arrow; Edge; Future; Electronics Omni Displays; Symmetry | $370 | 3.2% |

## Resources


1 About Light Emitting Diodes (LED). Retrieved Jun. 21, 2005, from Global Spec website: <http://light-emitting-diodes.globalspec.com/LearnMore/Optics_Optical_Components/Light_Sources/Light_Emitting_Diodes>.

2 About Photodiodes. Retrieved Jun. 23, 2005, from Global Spec website: <http://photodiode.globalspec.com/LearnMore/Optics_Optical_Components/Optoelectronics/Photodiodes.>.

3 Applications of Laser Diodes. Retrieved Jun. 23, 2005, from Wikipedia website: <http://en.wikipedia.org/wiki/Laser_diode#Applications_of_laser_diodes>.

4 Applications of Optocouplers. Retrieved Jul. 05, 2005, from Siemens website: <http://www.siemens.com>.

5 Coleman, J.J. (2003). Optoelectronics. In *Encyclopedia of Optical Engineering* (pp. 2015-2020). New York: Marcel Dekker, Inc.

6 Edwards, John. (2004, July 01). The global top 20. *Electronics Supply and Manufacturing*, Retrieved Jun 23, 2005, from website: <http://www.my-esm.com/showArticle.jhtml?articleID=22100763>.

7 Gordon, R. (2005). 2Q05 Semiconductor Forecast Update: Fears of aG00127762, Retrieved Jun 23, 2005, from Gartner Dataquest database.

8 Laser Diode. Retrieved Jun. 23, 2005, from Answers.com website: <http://www.answers.com/topic/laser-diode>.

9 LED Applications. Retrieved Jun. 23, 2005, from Wikipedia website: <http://en.wikipedia.org/wiki/LED#LED_applications>.

10 Muren, D. (2004, Jun 30). Tech Thursday: LED Me Not Into Temptation . *ID Fuel.* Retrieved Jun 23, 2005, from <http://www.idfuel.com/index.php?p=178&more=1>.

11 Optical Voice Link. Retrieved Jun. 23, 2005, from The Edgerton Center website: <http://web.mit.edu/Edgerton/www/Optical.html#>.

12 Optocoupler. (2005). Retrieved Jul 05, 2005, from Globalspec website: <http://optoelectronics.globalspec.com/LearnMore/Optics_Optical_Components/Optoelectronics/Optocouplers.>.

13 Optoelectronics In Today's World. Retrieved Jun. 23, 2005, from Optoelectronics Industry Development Association website: <http://www.oida.org/pdfs/oidainfobroch.pdf>.

14 Opto-isolator. (2005). Retrieved Jul 05, 2005, from Wikipedia website: <http://en.wikipedia.org/wiki/Optocoupler>.

15 P Glossary. Retrieved Jun. 23, 2005, from Aerotech Electronics – Fibre Optic Systems website: <http://www.aefos.com/html/glossary/p.htm>.

16 Photocells. Retrieved Jun. 23, 2005, from Perkinelmer website: <http://www.optoelectronics.perkinelmer.com/content/RelatedLinks/Photocells.pdf>.

17 Photocell Sensor. Retrieved Jun. 23, 2005, from Making Things website: <http://www.makingthings.com/teleo/cookbook/photocell.htm#overview>.

18 Photodiode. Retrieved Jun. 23, 2005, from Wikipedia website: <http://en.wikipedia.org/wiki/Photodiode>.

19 Photodiodes. Retrieved Jun. 23, 2005, from Perkinelmer website: <http://www.optoelectronics.perkinelmer.com/content/RelatedLinks/Photodiodes.pdf>.

20 Phototransistors. Retrieved Jun. 23, 2005, from Perkinelmer website: <http;//optoelectronics.perkinelmer.com/content/RelatedLinks/Phototransistors.pdf>.

21 Photovoltaics. Retrieved Jun. 23, 2005, from Arkansas Renewable Natural Energy website: <http://www.arkansasrenewableenergy.org/solar/pv.html#benefits>.

22 Soclof, S. (2005). Optoelectronic Devices. *The Engineering Handbook, 2nd Edition.* Retrieved Jun 23, 2005, from ENGnetBASE database.

23 Solar Cell. Retrieved Jun.21, 2005, from Wikipedia website: <http://en.wikipedia.org/wiki/Solar_cell>.

24 Typical Phototransistor and IRED applications. Retrieved Jun. 23, 2005, from Perkinelmer website: <http://optoelectronics.perkinelmer.com/content/RelatedLinks/PhototransistorTypicalApplications.pdf>.

25 What is Optoelectronics?. Retrieved Jun. 23, 2005, from Optoelectronics Industry Development Association website: <http://www.oida.org/whatisoe.html>.

# Processors

# 16

# Digital Signal Processors

A **digital signal processor (DSP)** is a specialized microprocessor designed specifically for converting signals from real world into digital data that can be analyzed. In this case, a signal means an electrical signal or stream of information carried by a wire or a radio wave that is processed in real time[4]. Digital signal processing can restore vintage jazz recordings to their original clarity, erase the static from long-distance phone lines and enable satellites to pick out global objects as small as a golf ball. In cars, DSPs create digital audio "surround sound" and are responsible for active suspension systems that adjust automatically to road conditions. In cellular phones, digital signal processors help squeeze more conversation onto crowded airwaves and can scramble signals to prevent eavesdroppers. In multimedia computers, digital signal processing enables business communication at the user's fingertips and professional audio sound in real time[5].

Formerly used in academic research and advanced military applications, digital signal processing has become a widely accessible commercial technology. DSPs were created to conduct processor mathematics that is extremely repetitive and data-intensive so that real-world signals like movement, sound and images can be represented. "The DSP was developed to be most efficient in handling the highly repetitive, data-intensive mathematics performed in processing data that represent real-world signals, such as motion, sound and images"[8]. In the last few years, a variety of high-performance, integrated DSP's have made digital signal processing technology easier and more affordable to use, particularly in low-end applications.

A DSP has four key components as seen in Figure 1[1]:

- **Program Memory:** Stores the programs the DSP will use to process data
- **Data Memory:** Stores the information to be processed
- **Compute Engine:** Performs the math processing, accessing the program from the Program Memory and the data from the Data Memory
- **Input/Output:** Serves a range of functions to connect to the outside world

**Figure 1:** Four Components of a Digital Signal Processor[1]

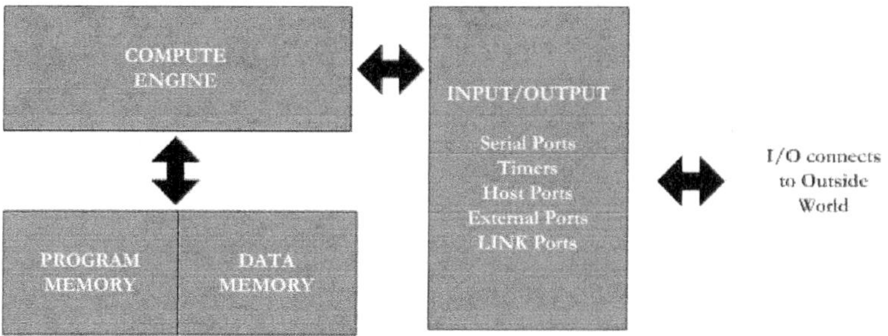


The most common type of digital signal processor is a **fixed point DSP** that operates on 16 or 32 bit words and performs strictly integer arithmetic; as the number of bits increase, the range also widens[7]. When comparing a fixed point DSP to a floating point, fixed point DSPs are cheaper and have a narrower dynamic range. The dynamic range is the spread of numbers that can be processed in the course of an application. In common signal processing applications, extra precision is not required placing an emphasis on speed. The fixed point variety is ideal for lower volume, lower cost applications including, but not limited to: audio systems, cell phones, digitized voices, or wireless communications. All of these applications are high volume and require a relatively narrow range of sound frequencies[7].

A **floating point DSP** operates on 32-40 bit words and performs either integer or real arithmetic. The floating point has wider dynamic range than fixed point, and it can manipulate more bits[7]. Floating point DSPs are generally used for image processing, 3-D graphics, scientific simulations, and medical applications, all of which require high precision and accuracy. The price of a fixed-point DSP is a tiny proportion of what a floating-point DSPs price is due to the large volumes of fixed-point DSP consumption and the increased complexity of the floating-point device. Therefore, floating-point DSPs are used by developers with low-volume applications that require very accurate mathmatics[2].

# The Product

## Applications and End Uses

Digital Signal Processors are used in audio signal processing (CD players, sound cards), video signal processing (hard disks, digital TV, video recorders, multimedia computers), and telecommunication devices (cell phones, modems, fax machines)[7]. An important application of DSP's is in **signal compression** and **decompression**. In CD systems, for example, the music recorded on the CD is in compressed form to increase storage capacity and must be decompressed for the recorded signal to be reproduced with high fidelity sound quality. Signal compression is used in digital cellular phones to allow a greater number of calls to be handled simultaneously within each local cell. When DSPs are used in signal compression technology, it allows people not only to talk to one another by telephone but also facilitates the capability to see one another on their computer monitors with only a conventional telephone line linking them together. The **architecture** of a DSP chip is designed to carry out such operations incredibly fast, processing up to tens of millions of data samples per second to provide real-time performance.

As the DSP architecture evolves, areas such as **Voice-over-Internet Protocol (VoIP)** will benefit from this new technology. The shift from traditional telephone networks to well-managed IP-based data networks is a growing trend that provides system efficiency and decreased calling costs. A DSP-based VoIP system or gateway makes a shift to data networks possible, serving as the bridge between the public switched telephone network (PSTN) and the packet network. VoIP gateways allow users to speak on traditional phones or send information over traditional fax machines as they bypass PSTN toll charges without a loss in quality[4].

## Product Features and User Benefits

**Figure 2:** DSP Features and Benefits

| Product Features | | User Benefits |
|---|---|---|
| Speed | ⇨ | Processes real world signals in real time to avoid discontinuities between inputs and outputs |
| Universal Series Bus (USB) connectable | ⇨ | Can be connected easily to peripheral devices by common USB cable |
| Standby Function | ⇨ | Can be kept on standby and ready to process without wasted excess power |
| Small Size | ⇨ | Gives designer cost and space flexibility when designing board architecture |
| Timing Capabilities | ⇨ | Can be placed on a timer to start, stop, or both based on user's specifications |
| Failure Detection | ⇨ | Can monitor system and initiate reset if a problem or failure is detected |
| Independent Functionality | ⇨ | Contains memory, processing, and input/output functionality all on one chip |

## Life Cycle

The DSP product life cycle begins with the introduction phase; at this point the supplier has announced the product and customers initiate designs. During the growth phase of a DSP's life cycle the majority of customers are using the product in prototype quantities and mass quantity production amounts will become available soon. The maturity phase of a DSP's product life cycle marks the point when most of the product's production has ramped and many customers are finding new technologies for their most recent designs. In the decline phase of a DSP there are not many new end-equipment designs that will use this technology and most activity is focused on long-term production for existing programs. The phase-out of a DSP is the stage where no new end-equipment designs even consider this technology, and the bulk of activity is focused on limited long-term production for continued end-equipment programs[6] and new designs with a more viable DSP solution.

**Figure 3:** Life Cycle Curve for DSP Technologies[6]

(Labels, in order, correspond with points from left to right on the Life Cycle Curve)

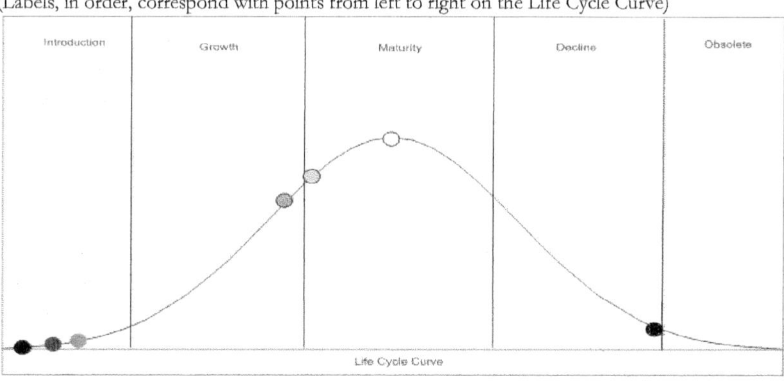


- Fixed Point w/ clock rates (CR) up to 1 GHz & 8000 Millions Instructions per Second (MIPS)
- Floating Point w/ CR up to 150 MHz & 150 MIPS; also has on-board Flash Memory
- Fixed Point w/ CR up to 300 MHz & 1800 MIPS
- Floating Point w/ CR up to 33 MHz & 16.67 MIPS
- Fixed Point w/ CR up to 66 MHz & 33 MIPS
- Floating Point w/ CR up to 50 MHz & 25 MIPS
- Fixed Point w/ CR up to 20 MHz & 5 MIPS

# The Market

## Market Trends

Figure 4 shows the DSP total available market for the years 2000 through a forecasted 2009.

**Figure 4:** DSP TAM Trend for 1999-2009 in $B(US)[8;9;11]

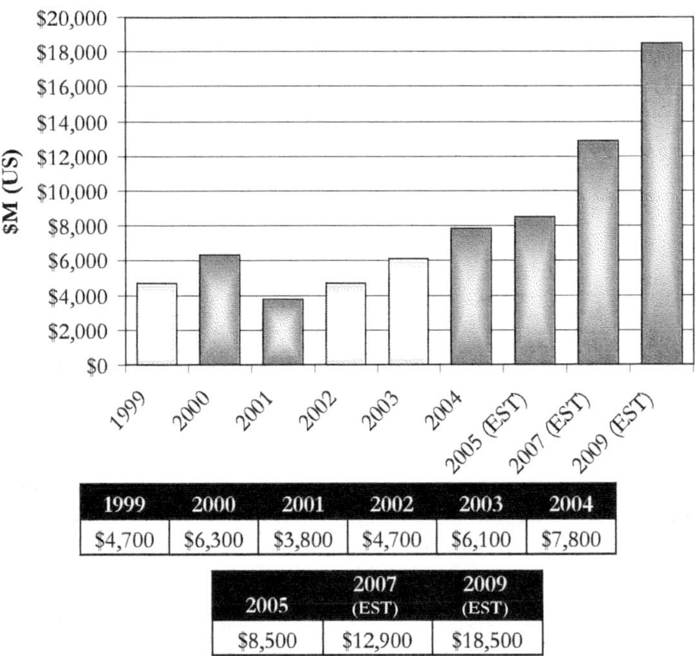


| 1999 | 2000 | 2001 | 2002 | 2003 | 2004 |
|---|---|---|---|---|---|
| $4,700 | $6,300 | $3,800 | $4,700 | $6,100 | $7,800 |

| 2005 | 2007 (EST) | 2009 (EST) |
|---|---|---|
| $8,500 | $12,900 | $18,500 |

## Regional Segmentation

With the recent manufacturing shift to China, the Asia Pacific region now accounts for over half of the world's DSP market[9].

**Figure 5:** DSP Regional TAM Segmentation for 2004[9]

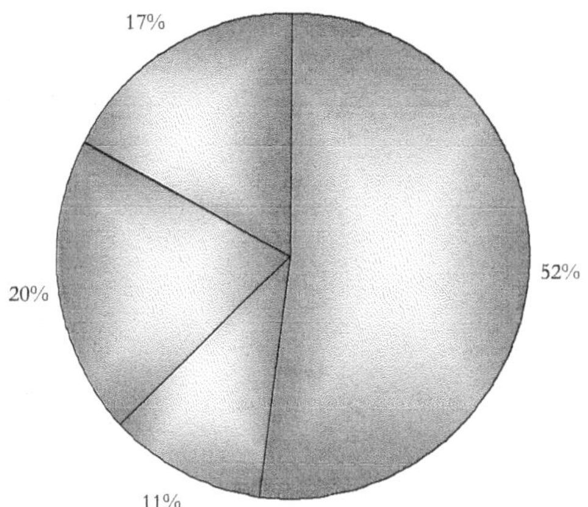


| DSP Regional Market Segmentation | |
|---|---|
| Asia Pacific | 52.0% |
| Europe | 20.0% |
| Japan | 17.3% |
| Americas | 10.7% |

## Market Segmentation

With a 33 percent growth rate in 2004, the wireless market accounts for almost three fourths of the DSP demand. This is due primarily to the increase in Wi-Fi "hot spots" and the growth in the VoIP sector[9]. There are five other areas that account for the DSP market along with the Wireless sector as shown in the figure below.

**Figure 5:** DSP Market Segmentation for 2004[9]

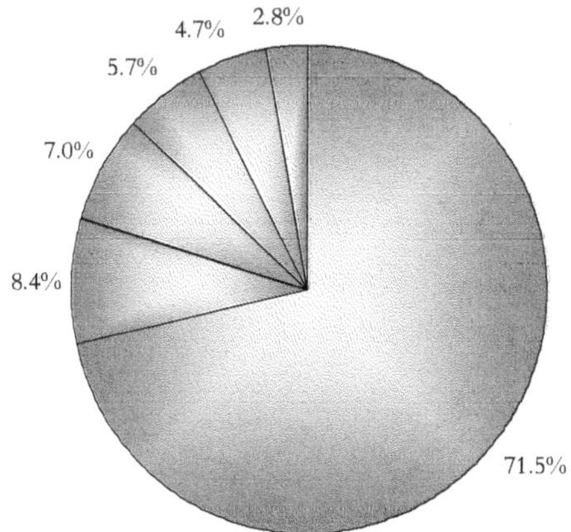


| Market Segmentation by Application | |
|---|---|
| Wireless | 71.5% |
| Consumer | 8.4% |
| Multi-purpose | 7.0% |
| Wired | 5.7% |
| Computer | 4.7% |
| Automotive | 2.8% |

# The Players

## Key Players

Texas Instruments has commanding market share with a approximately 50 percent of the worldwide market. The majority of the remaining market share is comprised of four other DSP manufacturers: Freescale, Agere, Analog Devices, and Philips Semiconductor.

**Figure 6:** Top Five DSP Manufacturers in 2004[9]

| | Supplier | Website | Distributors | Revenue $M(US) | Market Share % |
|---|---|---|---|---|---|
| 1 | Texas Instruments | www.ti.com | Arrow; Avnet; Digi-Key; Newark InOne | $30,300 | 50% |
| 2 | Freescale (Motorola) | www.freescale.com | Arrow; Avnet; Future; Premier Future | $12,580 | 13% |
| 3 | Agere | www.agere.com | Arrow; Avnet | $5,700 | 10% |
| 4 | Analog Devices | www.analog.com | Arrow; Avnet; Classic Components; RS Electronics | $2,600 | 8% |
| 5 | Philips Semiconductor | www.semiconductors.philips.com | All American; Avnet; Classic Components | $1,912 | 7% |

## References

1 A Beginner's Guide to DSP. (2005). Retrieved Jul. 13, 2005 from Analog Devices Web site: www.analog.com/processors/resources/beginnersGuide/index.html.

2 Allen, R. (2004). Converting floating-point applications to fixed-point. Retrieved Jul. 12, 2005 from Embedded Web site: http://www.embedded.com//showArticle.jhtml?articlID=47903200.

4 Digital Signal Processor. (2005). Retrieved Jul. 13, 2005 from Wikipedia Web site: http://en.wikipedia.org/wiki/Digital_signal_processor.

5 DSP. (2002). Retrieved Jul. 13, 2005 from Webopedia Web site: http://www.webopedia.com/TERM/d/dsp.html.

6 DSP Life Cycles. (). Retrieved Jul. 13, 2005 from TI Web site: http://focus.ti.com/docs/military/catalog/general/general.jhtml?templateId=5603&path=templatedata/cm/milgeneral/data/dsp_lifecycle&familyId=44

7 DSP Tutorial. (2001). Retrieved Jul. 13, 2005 from Wave Report Web site: http://www.wave-report.com/tutorials/DSP.htm

8 Starnes, T. (2002). Functions Distinguish Programmable Processors. Retrieved Jul. 12, 2005 from Garter Database.

9 Strauss, W. (2005). Forward Concepts' DSP/Wireless Market Bulletin. Retrieved Jul. 13, 2005 from Forward Concepts Web site: http://www.forwardconcepts.com/DSP%20Bulletin_2.9.05.pdf.

10 Strauss, W. (2004). DSP Stays the Course. Retrieved Jul. 13, 2005 from EETimes Web site: www.eetimes.com/showArticle.jhtml?articleID=21400223.

11 Walker, J. (2002). Semiconductor Packaging Forecast. Retrieved Jul. 13, 2005 from Gartner Database.

# 17 Microcontrollers

A **microcontroller** is a single-chip computer used to control some process or function. "Micro" suggests that the device is small, and "controller" tells you that the device might be used to control objects, processes, or events[1]. A microcontroller has an on-board program ROM and I/O (input/output) that can be programmed for various control functions. In general, any product or device that interacts with its user has a microcontroller buried inside it. The basic use for a microcontroller is to gather input from various sensors, process this input into a set of actions, and then use the output mechanisms on the microcontroller to do something useful. Some common uses for microcontrollers include: microwave keypads, remote controls, automobile components, and cell phones[10].

At one time, controllers were built exclusively from logic components, and were usually large, heavy boxes. After that, microprocessors were used, and the entire controller could fit on a small circuit board, which is still the common practice. Many controllers are powered by microprocessors. However, as the process of miniaturization continued, all of the components needed for a controller were built onto one chip. The one-chip computer, or microcontroller, was born. A microcontroller is a highly integrated chip that includes, on one chip, all or most of the parts needed for a controller[2].

A microcontroller is similar to the microprocessor inside a personal computer. Both microprocessors and microcontrollers contain a central processing unit, or CPU. The CPU executes instructions that perform the basic logic, math, and data-moving functions of a computer. To make a complete computer, a microprocessor requires memory for storing data and programs, and input/output (I/O) interfaces for connecting external devices, such as keyboards and displays.

In contrast to a microprocessor, a microcontroller is a single-chip computer because it contains memory and I/O interfaces in addition to the CPU. Because the amount of memory and interfaces that can fit on a single chip is limited, microcontrollers tend to be used in smaller systems that only require a few support components[1]. A microcontroller typically includes:

- CPU (central processing unit)
- RAM (random access memory)
- EPROM/PROM/ROM (erasable programmable read only memory)

- Serial and Parallel I/O (input/output)
- Timers
- Interrupt Controller

A typical microcontroller has bit manipulation instructions, easy and direct access to I/O, and quick and efficient interrupt processing. Microcontrollers are a "one-chip solution" that reduces parts count and design costs[2].

An **embedded controller** (or **embedded processor**) is a controller that is built into, or embedded, in a greater system that they help control[1]. Embedded controllers adhere to a philosophy similar to that of microcontrollers—high integration. By including many features necessary for the task at hand, an embedded controller (processor) can be a powerful yet cost-effective solution. However, while a microcontroller (almost by definition) is a computer on a chip, an embedded controller might need external components before it is considered a "computer"[2].

A microcontroller, or embedded controller, is a small computer that usually serves one purpose. For instance, the microcontroller inside a TV takes the input from a remote control and displays the data on the screen. A microcontroller is normally small and low cost, minimizing product size and expense. A microcontroller must sometimes be "ruggedized" to sustain harsh conditions. Automobile applications usually require these types of chips whereas chips inside a TV do not encounter harsh conditions[10].

Microcontrollers also come in a variety of capabilities. For example, a microwave keypad does not require a microcontroller as complex as one you would find in a cell phone. The most common microcontrollers are the 4-bit, 8-bit, 16-bit, and 32-bit chips. The demand for the 4-bit microcontroller is decreasing due to the increased advancement of technology but they are still used for some simple applications. Recently, the 64-bit chip has been used for some high-end applications[10].
Most microcontrollers today are based on the von Neumann architecture, which clearly defined the four basic components required for an embedded system. These include a CPU core, memory for the program (ROM or Flash memory) and memory for data (RAM), one or more timers (customizable ones and watchdog timers), as well as I/O lines to communicate with external peripherals and complementary resources—all this in a single integrated circuit. A traditional microprocessor requires all of these tasks to be handled by other chips. For example, RAM memory chips must be added. The amount of memory provided is more flexible in the traditional approach, but at least a few external memory chips must be provided, requiring additional connections to be made to pass the data back and forth to them[12].

For instance, a typical microcontroller has a built-in clock generator and a small amount of RAM and ROM (or EPROM or EEPROM), meaning that to make it work, all that is needed is some control software and a timing crystal.

Microcontrollers also have a variety of input/output devices, such as analog-to-digital converters, timers, UARTs, or specialized serial communications interfaces, such as IC serial peripheral interface, and controller area network. Often these integrated devices can be controlled by specialized processor instructions[12].

Some modern microcontrollers include a built-in, high-level programming language. BASIC is quite commonly used.

Microcontrollers trade away speed and flexibility to gain ease of equipment design and low cost. There's only so much room on the chip to include functionality, so for every I/O device or memory increase the microcontroller includes, some other circuitry has to be removed. Finally, some microcontroller architectures are available from many different vendors in so many varieties that they could rightly belong to a category of their own; chief among these are the 8051 and Z80 derivatives[12].

# The Product

## Applications and End Uses

You can find microcontrollers in all kinds of things. Any device that measures, stores, controls, calculates, or displays information is a candidate for putting a microcontroller inside. The largest single use for microcontrollers is in automobiles—just about every car manufactured today includes at least one microcontroller for engine control, and often more to control additional systems in the car, such as diagnostics and climate control. In computers and computer equipment, you can find microcontrollers inside keyboards, modems, printers, modems, disk drives, and other parts. They are found in appliances and consumer products, such as microwave ovens, refrigerators, television, VCRs, stereos, and cameras. In test equipment, microcontrollers make it easy to add features, such as the ability to store measurements, to create and store user routines, and to display messages and waveforms. Other applications include instrumentation and aerospace. In many items, more than one processor can be found[1].

Embedded processors and microcontrollers are used extensively in robotics. In this application, many specific tasks might be distributed among a large number of controllers in one system. Communications between each controller and a central, possibly more powerful controller (or micro/mini/mainframe), enable information to be processed by the central computer, or to be passed around to other controllers in the system[2].

A special application that microcontrollers are well suited for is data logging, such as monitoring and recording environmental parameters (temperature, humidity, rain). Small size, low power consumption, and flexibility make these devices ideal for unattended data monitoring and recording[2].

## Product Features and User Benefits

**Figure 1:** Features and User Benefits of Microcontrollers

| Product Features | | User Benefits |
|---|---|---|
| Fabrication Techniques – CMOS and PMP | ⇨ | CMOS (complementary metal oxide semiconductor)–less power usage, allows battery operation, high immunity to noise. PMP (post metal programming)–ROM can be programmed late in the production cycle. |
| Architectural Features | ⇨ | Von Neuman–can fetch both instruction and data CISC (complex instruction set computer)–allows programmer to use one instruction in place of many simpler ones. RISC (reduced instruction set computers)–smaller chip, pin count, low power consumption. SISC (specific instruction set computer)–saves space on a computer so more features can be added. |
| Advanced Memory Options | ⇨ | EPROM–faster, and permits more erase/write cycles. Static RAM–no limit to number of times written. Nonvolatile memory OTP–software protection |
| Power Management and Low Voltage | ⇨ | Longer battery life. |
| Watchdog Timer | ⇨ | Recovery from system problems. |
| Use of Digital Signal Processors (DSP) | ⇨ | Helps improve telephone quality voice, and marginal room for errors on compact disks. |

**Figure 2:** Common Microcontrollers and the Manufacturers[12]

| Manufacturer | Microcontrollers |
|---|---|
| **Atmel** | Atmel AT91 series (ARM THUMB architecture)<br>AT90 series – AVR (Atmel Norway design)<br>Atmel AT89 series (Intel 8051/MCS51 architecture)<br>MARC4 |
| **Cypress MicroSystems** | CY8C2xxxx (PSoC) |
| **Intel** | 8-bit<br>8XC42<br>MCS48<br>MCS51<br>8xC251<br>16-bit<br>MCS96<br>MXS296<br>32-bit: i960 |
| **Microchip** | 12-bit instruction PIC<br>14-bit instruction PIC<br>PIC16F84<br>16-bit instruction PIC |
| **Motorola** | 8-bit<br>68HC08 (CPU08)<br>68HC11 (CPU11)<br>16-bit<br>68HC16 (CPU16)<br>Motorola DSP56800<br>32-bit<br>Motorola 683XX (CPU32)<br>MPC 860 (PowerQUICC)<br>MPC 8240/8250<br>MPC 8540/8555/8560 |
| **National Semiconductor** | COP8<br>CR16 |
| **NEC** | 78K |
| **Parallax, Inc.** | BASIC Stamp |
| **Philips Semiconductors** | LPC2000<br>LPC900<br>LPC700 |
| **Renesas Tech. Corp.** | a joint venture of Hitachi and Mitsubishi. |
| **H8** | |
| **Holtek** | HT8 |
| **STMicroelectronics** | ST 62<br>ST 7 |
| **Texas Instruments** | TMS370<br>MSP430 |
| **Western Design Center** | 8-bit: W65C02-based μCs<br>16-bit: W65816-based μCs |
| **Ubicom** | SX-28 |
| **ZiLOG** | Z8<br>Z86E02 |

# The Market

## Market Trends[2]

Microcontrollers will be present and popular in the market until at least 2007, because they are indispensable for many applications and their price and ease of use makes them faster solutions than custom designs. The microcontroller unit that will decline the most is the 4-bit analog wireless, foreseen to drop at a compound annual decline rate of 23% through 2007. The 32-bit will be the highest growing unit, including 32-bit digital wireless, 32-bit DSL modem, and 32-bit monitor segments. Also, by 2007 it is predicted that the 16-bit hard drive and 8-bit infrared remote application segments will be in the top five of units shipped annually[4].

**Figure 3:** Worldwide Semiconductor Device Revenue Forecast[14]

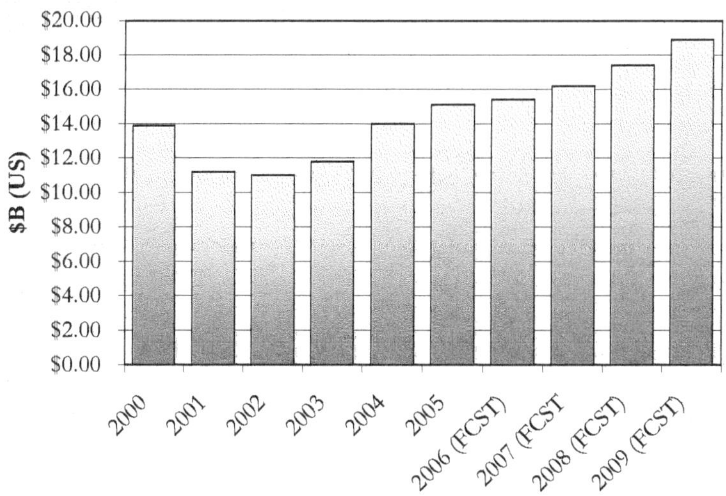


Source: iSuppli Corporation

| 2000 | 2001 | 2002 | 2003 | 2004 | 2005 |
|---|---|---|---|---|---|
| $31.20 | $23.50 | $23.00 | $25.20 | $28.80 | $31.90 |

| 2006 (FCST) | 2007 (FCST | 2008 (FCST) | 2009 (FCST) |
|---|---|---|---|
| **$15.4** | **$16.2** | **$17.4** | **$18.9** |

## Market Segmentation

Since 2000, the 32-bit and 64-bit microcontrollers have been introduced to the industry. Therefore, today there are five different types of microcontrollers on the market: 4-bit, 8-bit, 16-bit, 32-bit, and 64-bit. The 8-, 16-, and 32-bit microcontrollers are more popular in use than the 4- and 64-bit. Research firms generally predict a stable future for 8-bit micros. The 64-bit is only used in high end applications, meaning it is just at the beginning of its life cycle. A view of the market segmentation between 4-, 8-, 16-, and 32-bit microcontrollers is shown in Figure 4. Some research firms would argue that the 8-bit share is closer to 40% with most of the gain subtracted from the 16-bit slice[15].

**Figure 4:** Market Segmentation by Bit of Microcontrollers[15]

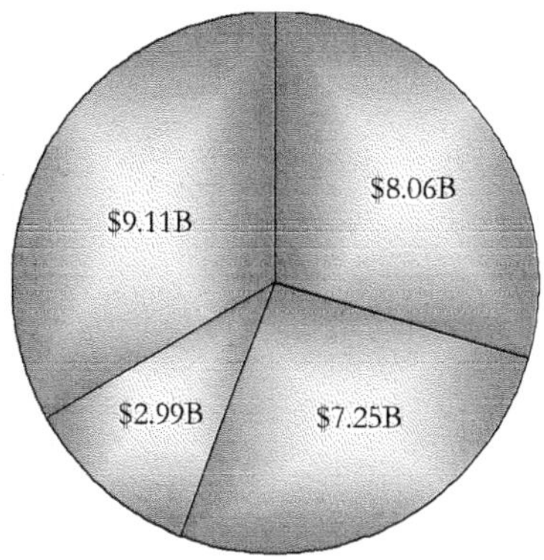


| 2005 Regional Segmentation | |
|---|---|
| Asia Pacific | $9.11 |
| Americas | $8.06 |
| Europe | $7.25 |
| Japan | $2.99 |

The industrial need for microcontrollers is growing rapidly. The automotive market is the most important driving force in the microcontroller market. Several microcontroller families were developed specifically for automotive applications and were subsequently modified to serve other embedded applications. Consumer applications in homes are also increasing. New applications in remote metering and security, such as electronically-controlled window locks, are creating new, potentially high-volume markets as well[15].

# The Players

## Key Players

**Figure 5:** Ranked Manufacturers of Microcontrollers in 2001[5]

| | Supplier | Website | Distributors | Revenue $M (US) | Market Share % |
|---|---|---|---|---|---|
| 1 | Freescale (Motorola) | www.freescale.com | Arrow; Avnet; Digi-Key; Future; Newark InOne; Richardson; Rochester | $550 | 20.4% |
| 2 | Infineon Technologies | www.infineon.com | ACG; Arrow; Avnet; Digi-Key; Future; Richardson | $452 | 16.7% |
| 3 | Renesas Technology (Hitachi) | www.renesas.com | All American; Arrow; Avnet; Digi-Key; NuHorizons | $307 | 11.4% |
| 4 | ST Microelectronics | www.st.com | Arrow; Avnet; Digi-Key; Mouser; NuHorizons | $254 | 9.4% |
| 5 | NEC Electronics | www.nec.com | Arrow; Edge; Future; Jaco; Symmetry | $225 | 8.3% |

| | Supplier | Website | Distributors | Revenue $M (US) | Market Share % |
|---|---|---|---|---|---|
| 6 | Microchip Technology | www.microchip.com | Arrow; Chip Supply; Future; Digi-Key; Mouser; Newark InOne | $171 | 6.3% |
| 7 | Philips Semiconductor | www.semiconductors.philips.com | Arrow; Avnet; Digi-Key; Future | $154 | 5.7% |
| 8 | Atmel | www.atmel.com | All American; Arrow; Avnet; Jan Devices | $114 | 4.2% |
| 9 | Texas Instruments (TI) | www.ti.com | Arrow; Avnet; Newark InOne; Rochester | $86 | 3.2% |
| 10 | Mitsubishi | www.mitsuishi.com | | $85 | 3.1% |

## References

1 Axelson, Jan. (1997). *The Microcontroller Idea Book; Circuits Programs and Applications featuring the 8052-BASIC Microcontroller.* Madison, WI: Lakeview Research. Retrieved Jun.14, 2005 from website: <http://www.lvr.com/files/mibch1.pdf#search='what%20is%20a%20microcontroller>.

2 Fischer, Kirsten. (2003). Over US$7.4 Billion Microcontrollers to be Shipped by 2007. Retrieved Jun. 14, 2005, from In-Stat Press Releases website: <http://instat.com/press.asp?ID=795&sku=IN0301097DE>.

3 Giovino, Bill. (2002). Microcontroller FAQ. Retrieved Jun. 14, 2005, from Microcontroller.com website: <http://www.microcontroller.com/ embedded/references/faqs/microcontroller-faq.htm>.

4 Gordon, Richard. (2004). Semiconductor Market Will Grow 25 Percent in 2004(G00120997), Retrieved Jun 14, 2005, from Gartner, Inc. database.

5 Hirst, C. (2002). EMEA Microcontroller Market Review, 2001 (Executive Summary)(SEMC-WW-EX-0169), Retrieved Jun 14, 2005, from Gartner, Inc. database.

6 Levine, Bernard. (2002, Dec 2). U.s. DTAM Dreary - Distribution Trends 2003 - Distribution Total Available Market - Industry Overview. *Electronic News*, Retrieved Jun 14, 2005, from <http://www.findarticles.com/p/articles/mi_m0EKF/is_49_48/ai_94982467>.

7 M683XX Overview. Retrieved Jun. 10, 2005, from Freescale Semiconductor website: <http://www.freescale.com/webapp/sps/site/overview.jsp?nodeId=0162462LcRQmlG>.

8 "Microchip Technology Unveils 16- bit dsPIC Family" 17 Nov 2000, Retrieved 18 Oct 2004 from: <http://www.microchip.com>.

9 Microcontrollers. Retrieved Jun. 10, 2005, from How Stuff Works website: <http://www.electronics.howstuffworks.com/microcontroller.htm.com>.

10 Microcontrollers. Retrieved Jun. 14, 2005, from NEC Online website: <http://www.necelam.com/microcontrollers/home.php>.

11 Microcontrollers. Retrieved Jun. 14, 2005, from Wikipedia website: <http://en.wikipedia.org/wiki/Microcontroller>.

12 MPU and MCU. Retrieved Jun. 14, 2005, from Renesas website: <http://america.renesas.com/fmwk.jsp?cnt=mpumcu_category_landing.jsp&fp=/products/mpumcu/>.

13 Nadamuni, Daya. Embedded Systems Development Trends: Asia (2004, Mar 01). *Electronic Engineering Times (Asia) and Gartner Dataquest*, Retrieved Jun 14, 2005, from <http://www.eetasia.com/ARTP_880030870_0.HTM>.

14 Olsson, M. (2002). Trends in Semiconductor Market, 2002(SEMC-WW-EX-0143), Retrieved Jun 14, 2005, from Gartner Inc, database.

15 Shandle, Jack. (2004, Aug 5). More for less: stable future for 8-bit microcontrollers. *TechOnLine*, Retrieved Jun 14, 2005, from <http://www.techonline.com/community/ed_resource/feature_article/36930?print>.

16 Solomon, Sandbourn and Pecht, . "Electronic Part Life Cycle Concepts and Obsolescence Forecasting." IEEE Trans. on Components and Packaging Technologies. Retrieved 10 Jun. 2005 from: <http://www.enme.umd.edu/ESCML/Papers/ObsCPMT.pdf>.

17 Spiegel, Rob. "India Preps for Manufacturing Boom" *Electronic News*, 9 Mar 2004. Retrieved 12 Oct 2004. from: <http://www.reed-electronics.com>.

18 Starnes, Tom. (2004). Stabilized Economies Allow Microcomponent Market to Grow Again(SEMC-WW-DP-0370), Retrieved Jun 14, 2005, from Gartner Inc, database.

19 Sullivan, Laurie. "Distribution" 18 Aug 2003 Retrieved 12 Oct 2004.

# 18 Microprocessors

A **microprocessor** is an integrated circuit that performs the bulk of the processing in most electronic products. In simpler terms, any microprocessor can be considered the brain of the system. Also referred to as the central processing unit (CPU), the microprocessor is a complete computation engine contained on a single chip. The first microprocessor (Intel 4004), introduced in 1971, did not contain much computing power and could only process 4 bits at a time. The first chip to be placed into a home computer, the Intel 8080, doubled the processing from 4 bits to 8 bits and increased the computing speed of the component. In 1979, Intel introduced the 8088 chip, which was incorporated into the IBM personal computer. Over time, Intel has continued to make improvements to the architecture of the 8088 chip. The newest chip on the market, the Pentium 4, executes at a rate of 5,000 times faster than the original 8088 chip[1]. While Intel dominates the microprocessor market, AMD, Motorola, and others also manufacture microprocessor products.

A microprocessor has three main functions:

1.) Computing mathematical functions using its **arithmetic/logic unit** (**ALU**).
2.) Moving data from one memory location to another.
3.) Making decisions while moving from one instruction to another.

The performance of the chip is largely related to the number of transistors available. More transistors enable an architectural feature called **pipelining** to be used, in which instruction execution can be overlapped. In this architecture, multiple instructions in various stages can be executed simultaneously.

A microprocessor contains several key components that facilitate its operation. The **address bus** and **data bus** send and receive addresses to and from the memory of the system. The **read (RD)** and **write (WR) lines** tell the memory whether it wants to set or receive an address location. Finally, the **clock line** lets a clock pulse sequence the processor, while the reset line allows the counter to reset and restart an execution[1]. The lines and busses of the processor connect to either read only memory (ROM), random access memory (RAM), or Flash devices. ROM is programmed with pre-set bytes. RAM also contains bytes of information. Unlike ROM, RAM can read or write to those bytes depending upon whether the RD or WR line is signaled. The drawback to RAM is that once the power is removed, it loses all its information, thus providing a need for a non-volatile memory like ROM, PROM, EPROM or EEPROM.

The three basic characteristics that distinguish different types of microprocessors are described in Figure 1.

**Figure 1:** Three Characteristics of Microprocessors[3]

| Characteristic | Definition |
|---|---|
| Instruction Set | The programmed instructions the chip can execute. |
| Bandwidth | The number of bits that are processed in a single instruction. |
| Clock Speed | Determines how many instructions per second the chip can execute, given in megahertz (MHz). |

# The Product

## Applications and End Uses

The single largest application of the microprocessor is as the CPU in a computer. In large computers, microprocessors are used as smart controllers for graphics displays, storage devices, and high-speed printers. The microprocessor can be used to control almost any electromechanical device. The hand-held electronic calculator has dropped dramatically in price due to the microprocessor. The digital wristwatch and electronic games are two other end-applications for the microprocessor. They can be seen in common household appliances, such as the microwave oven and VCR. In the automotive industry, they are used to regulate gasoline consumption and antilock brakes. Even the security industry uses microprocessors to monitor alarm systems. Finally, among other defense applications, microprocessors are used to operate automatic tracking and targeting systems in aircraft, tanks, and missiles and also to control radar arrays that track and identify aircraft[5].

## Product Features and Benefits

**Figure 2:** Product Features and User Benefits of Microprocessors

| Product Features | | User Benefits |
|---|---|---|
| High Speed Computations | ⇨ | ALU allows chip to process arithmetic and logic functions at high speeds. |
| Multiple Instructions | ⇨ | Pipeline architecture allows microprocessors to execute multiple instructions simultaneously. |

| | | |
|---|---|---|
| Backward Compatible | ⇨ | Can run programs written for a less powerful microprocessor, eliminating the need to re-write all microprocessor code as systems are upgraded. |
| Low Power Consumption | ⇨ | As microprocessors get smaller, the amount of power that they consume dramatically reduces. This allows users to run more powerful applications with cost-efficient power supplies. |
| 32- and 64-Bit Bus Bandwidth | ⇨ | Ability to process more bits per single instruction. Facilitates faster computation. |
| Gigahertz (GHz) Clock Speed | ⇨ | Can process hundreds, thousands, and millions of instructions per second. |

The following table compares and contrasts several important aspects found on some of the more popular chips in the market today. Over time, processor engineers have been able to add more transistors to their designs resulting in faster CPU clock speeds[3]. The **cache** refers to the collection of data that duplicates original values stored elsewhere or that has been computed earlier. This is typically used when the original data takes too much time to fetch or compute as compared to using the cache.

**Figure 3:** Important Aspects of Popular Microprocessors[3]

| Microprocessor | Transistors | CPU Speed | L2 Cache | Front-Side Bus Speed |
|---|---|---|---|---|
| **Celeron** | 7,500,000 | 1.06 GHz - 2 GHz | 256 KB, full speed | 133 MHz and 400 MHz |
| **Pentium II** | 7,500,000 | 233 MHz - 450 MHz | 512 KB, half speed | 100 MHz |
| **Pentium III** | 9,500,000 | 450 MHz - 1 GHz | 256 KB, full speed | 133 MHz |
| **Pentium III Xeon** | 28,100,000 | 500 MHz - 1 GHz | 256 KB - 2 MB, full speed | 100 MHz |
| **Pentium 4** | 55,000,000 | 1.4 GHz - 3.4 GHz | 256 KB, full speed | 800 MHz |
| **K6-II** | 9,300,000 | 500 MHz - 550 MHz | N/A | 100 MHz |
| **K6-III** | 21,300,000 | 400 MHz - 450 MHz | 256 KB, full speed | 100 MHz |
| **Athlon (K7)** | 22,000,000 | 850 MHz - 1.2 GHz | 256 KB, full speed | 200 MHz and 266 MHz |

| Athlon XP | 37,500,000 | 1.67 GHz | 384 KB, full speed | 266 MHz |
|---|---|---|---|---|
| Duron | N/A | 700-800 MHz | 64 KB, full speed | 200 MHz |
| PowerPC G3 | 6,500,000 | 233 MHz - 333 MHz | 512 KB, 1 MB, half speed | 100 MHz |
| PowerPC G4 | 10,500,000 | 400 MHz - 800 MHz | 1 MB, half speed | 100 MHz |
| Athlon 64 | 105,900,000 | 800 MHz | 1 MB, half speed | 1.6 GHz |
| Power PC G5 | 58,000,000 | 2.5GHz | 512 KB | 900MHz - 1.25GHz |

## Life Cycle

**Figure 4**: Life Cycle Curve for Intel Microprocessor

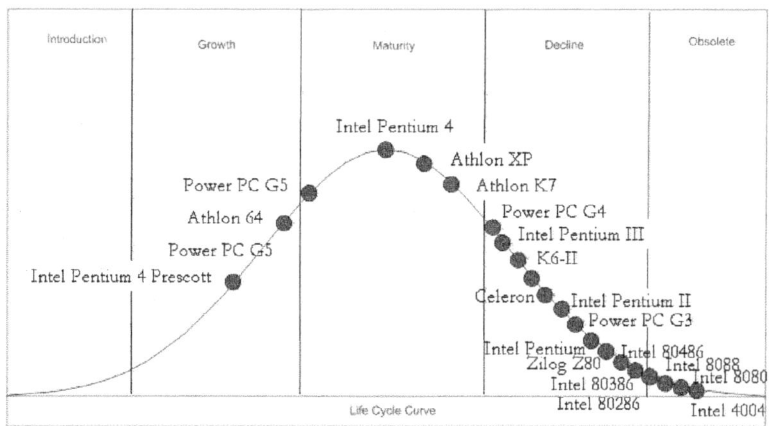


## Product Trends

The common trend for microprocessor design has been a full 32-bit ALU with "fast floating point processors" built in and created in pipeline form to execute multiple instructions. However, the newest microprocessor trend has been the 64-bit ALU. Also, customers are beginning to see special instruction microprocessors that will make specific applications very efficient, and the addition of hardware virtual memory support and L1 caching on the processor chip. These trends will increase the transistor count to the multi-million and they can execute around one billion instructions per second[1].

# The Market

## Market Trends

**Figure 5:** Total Available Market for MOS Microprocessors

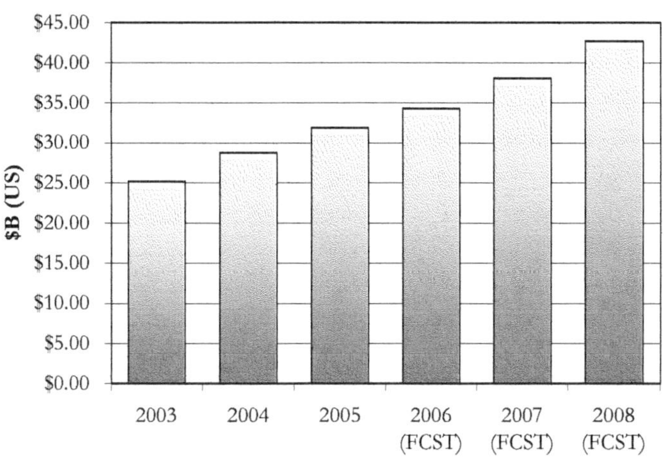


Source: iSuppli Corporation

| 2003 | 2004 | 2005 | 2006 (FCST) | 2007 (FCST) | 2008 (FCST) |
|---|---|---|---|---|---|
| $25.2 | $28.8 | $31.9 | $34.3 | $38.1 | $42.7 |

## Regional Segmentation

**Figure 6:** Regional TAM Segmentation Estimate for 2005

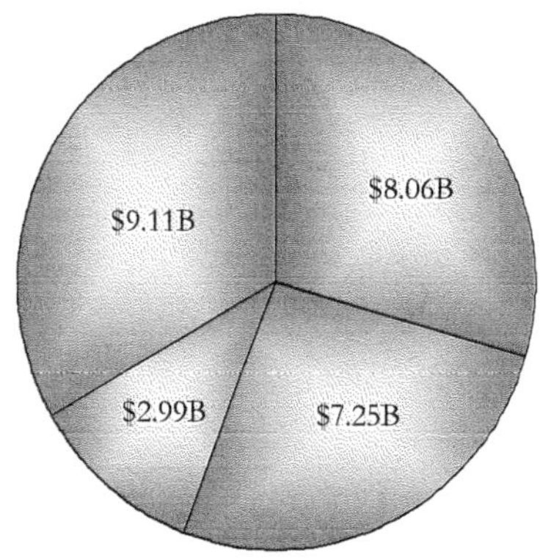


| 2005 Regional Segmentation | |
|---|---|
| Asia Pacific | $9.11 |
| Americas | $8.06 |
| Europe | $7.25 |
| Japan | $2.99 |

Source: WSTS provided by SIA

## Market Segmentation

**Figure 7:** Microprocessor Market Segmentation Estimation for 2005[2]

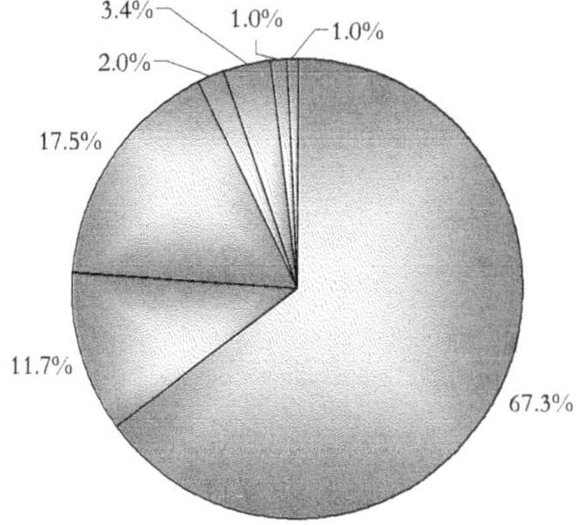


| Microprocessor Market Segmentation | |
|---|---|
| Telecom | 67.3% |
| Retail | 11.7% |
| Government | 17.5% |
| Transportation | 2.0% |
| Pay TV | 3.4% |
| Corporate Security | 1.0% |
| Others | 1.0% |

**Figure 8:** Microprocessor usage by Computer type for 2004[6]

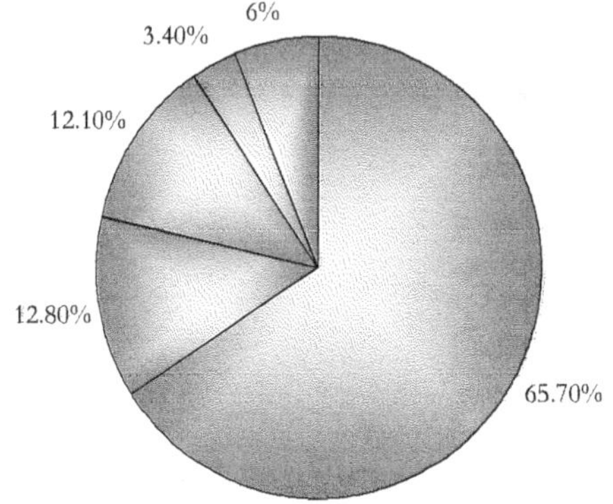


| Microprocessor Segmentation by Computer Type | |
|---|---|
| Desktops | 65.70% |
| Mobile | 12.80% |
| Mainframe | 12.10% |
| Workstation | 3.40% |
| Other | 6% |

# The Players

## Key Players

**Figure 9:** Top 80% of Microprocessor Suppliers of 2005[4]

| | Supplier | Website | Distributors | Market Share % |
|---|---|---|---|---|
| 1 | Intel | www.intel.com | Arrow<br>Avnet<br>Classic Components | 81.7% |
| 2 | AMD | www.amd.com | Arrow<br>Avnet<br>Premier Farnell | 16.9% |
| 3 | Others including but not limited to Freescale, Texas Instruments, VIA technologies, and Transmeta Corp. | | | 1.4% |

## References


1 Brain, M. (n.d.) How Microprocessors Work. Retrieved Aug. 2, 2005 from How Stuff Works website: <http://computer.howstuffworks.com/microprocessor.htm>.

2 Global smart card volumes by units and service type forecast for 2005. (2004). *Electronic Payments International* (ISSN: 0954-0393). Retrieved Aug. 2, 2005 from TableBase database.

3 Microprocessor. (2005). Retrieved Aug. 2, 2005 from Webopedia website: <http://www.webopedia.com/TERM/m/microprocessor.html>.

4 Shilov, A. (2005). Retrieved Aug. 2, 2005 from X-bit labs website: <http://xbitlabs.com/news/cpu/display/20050602023614>.

5 Slater, M. (1992). A guide to RISC Microprocessors. *Columbia University Press.* Retrieved Aug. 2, 2005 from the Free Dictionary website: <http://columbia.thefreedictionary.com/Microprocessors>.

6 United States microprocessor usage by computer type in percentages for 2004. (2004). *Purchasing* (ISSN: 0033-4448). Retrieved Aug. 2, 2005 from TableBase database.

# Interconnect, Passive, & Electromechanical

# Interconnect

# 19 Connectors

The term connector applies to virtually *any* product that connects devices. For the scope of this primer, the term **connector** will be applied to the specific range of electronic products whose sole purpose is to connect devices to one another. This connection is usually established through coupling of male and female connectors. Male connectors contain exposed pins or other form of contacts while female connectors contain holes, or sockets, in which the exposed pins are meant to be inserted. Male/female connectors are used in cases where convenience or manufacturing guidelines call for it.

**Circular connectors** are used for external device interfacing. They come in many sizes, differentiated by the number of contacts (pins and sockets) and contact diameter. In North America, the standard for contact diameter is the American Wire Gauge (AWG) standard[1].

**PCB connectors** mount onto Printed Circuit Boards (PCB); they use pins on the connector, they "mate" with holes on the PCB, and together they are then soldered. PCB connectors are used in cases where PCBs, such as boards and cards, are used. Some commonly found PCB connectors are Peripheral Component Interconnect (PCI), Deutsches Institut für Normung (DIN), Accelerated Graphics Port (AGP), and Personal Computer Memory Card International Association (PCMCIA).

**Rectangular connectors** comprise a large group of electronic connectors. Common rectangular connectors are the smaller Universal Serial Bus (USB) and IEEE 1394 Firewire® connectors[5].

**Telephone/telecom connectors** are types of modular connectors, meaning that they are made of "plug-in" units consisting of multiple, parallel contacts and can be added together to increase the size or capabilities of the system[4]. **Registered Jack (RJ) connectors** are a common type of telephone/telecom connector.

**Fiber optic connectors** enable the interconnection of separate fiber systems through fiber cables. Since fiber cables and systems rely on optical transmission through glass strands, the allowance for error in connecting systems is extremely low. Many standards for fiber system installation and manufacturing have been created because of this sensitivity and need for precision. Fiber optic connectors consist of multiple **ferrules** and an aligning sleeve to properly align the two optical surfaces, and a coupling sleeve to facilitate a secure connection[14].

# The Product

## Applications and End Uses

Connectors are used in almost every electronics market segment. Figure 1 shows a few of the segments with common examples of how connectors are used. Common end-use applications for circular connectors are industrial equipment and machines, computers and peripherals, medical equipment and audio and video equipment[1]. One of the larger end-use applications for rectangular connectors is PCs and peripherals[5]. RJ connectors are used in voice and data systems[4]. Fiber optic connectors are widely used in datacom, telecom, networking, security, and military applications[14].

**Figure 1:** Applications for Connector Products

| Segment | Consumer Application |
|---|---|
| Computers | mainframes, PCs, printers, modems |
| Consumer Appliances | dishwashers, dryers, washing machines, refrigerators, ranges |
| Telecom | data communication, telephones, radio and television |
| Aerospace | aircraft and space applications |
| Automotive | cars, trucks, farm equipment |

## Product Features and Benefits

**Figure 2:** Features and Corresponding User Benefits for PCB Connectors

| Product Features | | User Benefits |
|---|---|---|
| Kinked Pins | ⇨ | Maintains the connector securely in the PCB even allowing for the weight of cable during the solder process. "Bent" pins allow the assembly/solder process to continue without worry that the connector may slip from the board[12]. |
| Polarized Molding | ⇨ | Allows connectors to properly mate in only one way, assuring proper connection[13]. |
| Threaded Fasteners | ⇨ | PCB connector is held to other connectors with screws or nuts, insuring that connectors maintain a proper connection. |

**Figure 3:** Product Features and User Benefits of Fiber Optic Connectors

| Product Features | | User Benefits |
|---|---|---|
| Push/Pull Coupling | ⇒ | Connectors remain properly connected until the coupling is pushed in a specific direction, freeing the opposite connector, insuring that connectors stay connected even through rough movement[14]. |
| Low-Profile Latching System | ⇒ | Allows for smooth mating and de-mating of a connector, resulting in less risk of damage or improper connection or disconnection[6]. |
| Automatic Internal Shutter | ⇒ | Blocks out stray light, maintaining signal accuracy, and restricts contact between fiber core and ferrule endface, preventing damage[6]. |

**Figure 4:** Product Features and User Benefits of Coaxial Connectors

| Product Features | | User Benefits |
|---|---|---|
| Snap-In Connection | ⇒ | Easy connection, signaled by parts snapping together, ensuring a proper connection is established[10]. |
| Sequenced Mating | ⇒ | Mating sequence touches outer contacts before center contacts, reducing female center contact damage during mating, resulting in longer product life. |
| Extended Coupling | ⇒ | Plug coupling nut extends beyond the male center contact, protecting the contact while not mated. |

## Life Cycle

Registered jack systems, and therefore RJ connectors, have been in use for many years and have held their position at the peak of maturity. With a few exceptions, almost every telephone and computer networking system uses a form of RJ connector: RJ-45 for most computer networks and RJ-11 for a majority of telephone systems. With limited connection alternatives in the market in terms of modular capability and ease of use, the RJ classification of connectors will likely hold its position for many more years.

Universal serial bus connectors have, for lack of a better term, taken the world by storm; new devices everywhere use USB interconnect capabilities. Wireless phones are now able to use a USB adapter not only to facilitate data exchange to more efficiently organize contact information and calendar events, but to even act as a phone charger when an electrical outlet may not be readily accessible. In addition to wireless phones, USB has also made a huge impact in the computer peripheral market. Devices such as controllers, printers, digital imaging equipment, and even hard disk drives are now instantly available to PCs through a single USB connection. While too broad to discuss in detail, the impact and utility that USB interconnection offers holds it at peak maturity.

Accelerated graphic port connectors are used heavily in the PC market segment. Once an optional replacement for a PCI connector, AGP connectors for mainboards have become more of a norm than an option. This is due to the ever-increasing capabilities and demands of video-centric computing fields, such as three-dimensional imaging abilities and high-bandwidth requirements for PC applications. AGP connectors are secured, for the time being, at peak maturity; they're common enough to be somewhat of a commoditized product, yet so heavily used as to not begin the downhill trek toward obsolescence.

PCI connectors, while replaced by AGP connectors for video hardware use, are still widely used. PCI connectors are what connect a motherboard on a PC with sound cards, networking cards, various extensions, and more. The capabilities of PCI connectors are still adequate for these functions; however, with the inevitable technological advancements of the future, PCI will likely see a further steady decline in usage.

With the introduction of USB, D-subminiature and DIN connectors have lost market momentum. Keyboards and other peripheral devices that once drove widespread use of D-subminiature and DIN connectors are now being replaced by simpler, more efficient connector systems such as USB and IEEE 1394 Firewire™ systems.

On the maturing side, fiber optic connectors will see an increase in use as systems become easier to use and connectors and networking become more standardized and efficient. Currently fiber optic connectors are used in military, networking, datacom/telecom, and security systems applications[14].

---

™ Firewire is a trademark of Apple Computer, Inc. filed in 1993

**Figure 5:** Life Cycle Curve for Select Connectors

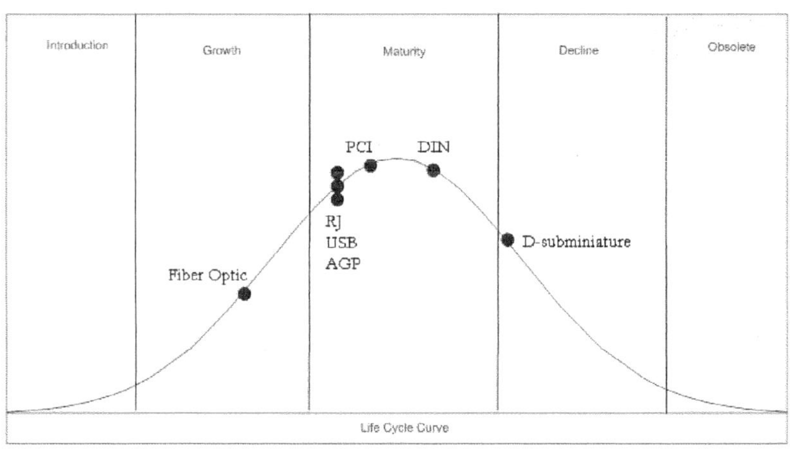


# The Market

## Market Trends

**Figure 6:** Past and Forecasted Worldwide TAM ($M)[7; 8]

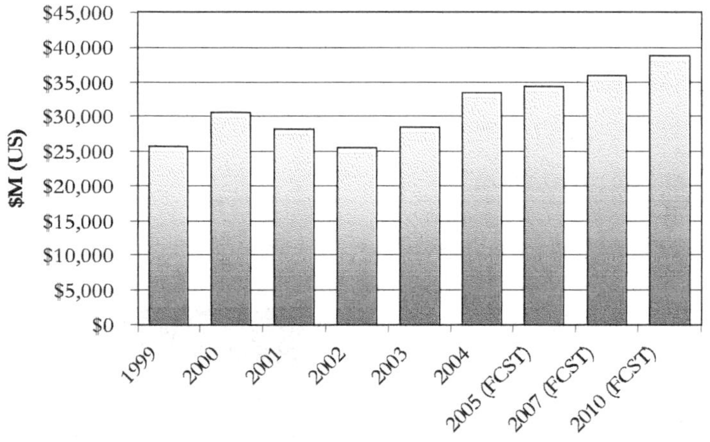


| 1999 | 2000 | 2001 | 2002 | 2003 | 2004 |
|---|---|---|---|---|---|
| $25,693 | $30,529 | $28,206 | $25,493 | $28,352 | $33,419 |

| 2005 (FCST) | 2007 (FCST) | 2010 (FCST) |
|---|---|---|
| $34,254 | $35,989 | $38,756 |

## Regional Segmentation

The worldwide connector market took a noticeable decline from 2000 to 2002 but regained market momentum beginning in 2003 and continuing into 2005 (as of time of publication). China's increase in shipment dollars marks an impressive show of the country's ability to swiftly and surely influence the world electronic connector market, surpassing the balance of the Asia-Pacific region in 2004.

**Figure 7:** Regional Connector Market Segmentation for 2004[8]

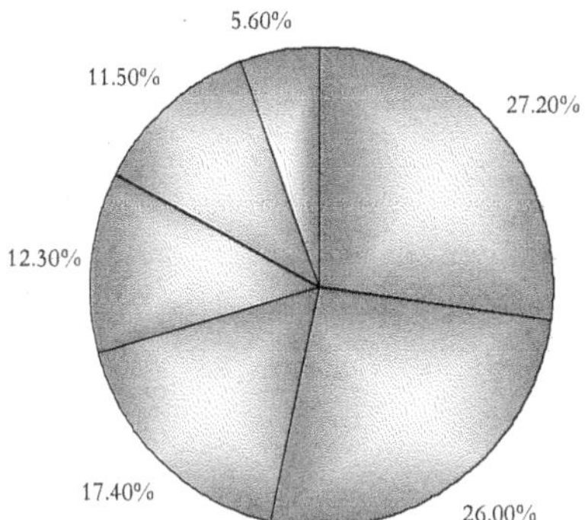


| Regional Connector Market Segmentation for 2004 | |
|---|---|
| North America | 27.2% |
| Europe | 26.0% |
| Japan | 17.4% |
| China | 12.3% |
| Rest of Asia-Pacific | 11.5% |
| ROW | 5.6% |

With the entry of China in the world connector market and the country's astounding growth, this return has been hastened. Forecasted values using a Compound Annual Growth Rate (CAGR) over the past five years puts the North American market at a decline while Europe and China experience the highest growth rates.

## Market Segmentation

**Figure 8:** Total Connector Market Segmentation for 2004[8]

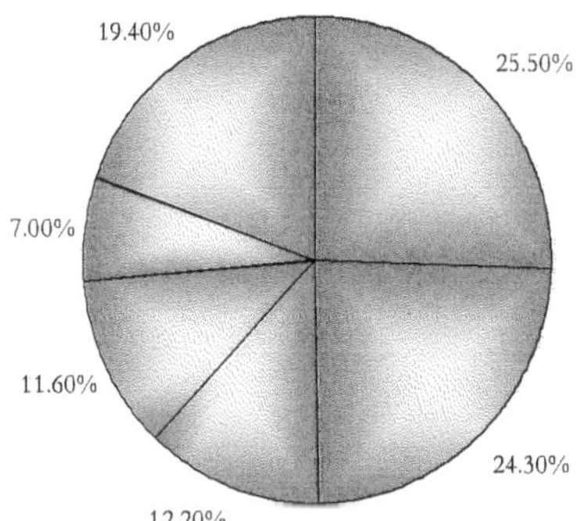


| Total Connector Market Segmentation | |
|---|---|
| Automotive | 25.5% |
| Computers & Peripherals | 24.3% |
| Telecom/Datacom | 12.2% |
| Industrial | 11.6% |
| Mil/Aero | 7.0% |
| All Others | 19.4% |

# The Players

## Key Players

The electronic connector industry varies somewhat from "traditional" numbers in that the top ten electronic connector manufacturers only contribute 57.7% of worldwide shipments. Tyco Electronics, however, currently has a lead on the rest of the market, taking up 21.5% of worldwide shipments. The next two leaders are Molex, Inc. with 7.7% of worldwide share and FCI with 5.7% .

**Figure 9:** Worldwide Top Three Electronic Connector Suppliers[9]

| | Supplier | Website | Distributors | Revenue $M (US) | Market Share % |
|---|---|---|---|---|---|
| 1 | Tyco Electronics | www.tycoelectronics.com | Arrow; Avnet; Premier Farnell; TTI | $4,962 | 21.5% |
| 2 | Molex, Inc. | www.molex.com | Avnet; RS Electronics; Sager; TTI | $1,789 | 7.7% |
| 3 | FCI | www.fciconnect.com | Avnet; Future; Premier Farnell; TTI | $1,321 | 5.7% |

## References


1 About Circular Connectors. (2005). Retrieved Feb. 20, 2005 from Globalspec website: <http://connectors.globalspec.com/LearnMore/Electrical_Electronic_Components/Connectors/Circular_Connectors>.

2 About Circular Connectors. (2005). Retrieved Feb. 21, 2005 from Globalspec website: <http://connectors.globalspec.com/LearnMore/Electrical_electronic_components/connectors/circular_connectors>.

3 About DIN Connectors (2005). Retrieved Feb 20, 2005 from Globalspec website: <http://connectors.globalspec.com/LearnMore/Electrical_Electronic_Components/Connectors/DIN_Connectors>.

4 About Modular Connectors and RJ Connectors. (2005). Retrieved Feb. 20, 2005 from Globalspec website: <http://connectors.globalspec.com/LearnMore/Electrical_Electronic_Components/Connectors/RJ_Type_Modular_Connectors

5 About USB Connectors. (2005). Retrieved Feb. 20 2005 from Globalspec website:<http://connectors.globalspec.com/LearnMore/Electrical_Electronic_Components/Connectors/USB_Connectors>.

6 Amphenol-Fiber Optic Products. (1997). Retrieved Feb. 20, 2005 from Amphenol Fiber Optics website: <http://www.amphenol-fiberoptics.com>.

7 Bishop & Associates. (2004). Connector Market Handbook. St. Charles, IL: Bishop & Associates.

8 Bishop & Associates. (2005). Connector Market Handbook. St. Charles, IL: Bishop & Associates.

9 Bishop & Associates. (2004). The Worlds Top 100 Connector Manufacturers. St. Charles, IL: Bishop & Associates.

10 Conec. (1996). Retrieved Feb. 15, 2005 from website: <http://www.conec.com>.

11 Connectors. (2005) Retrieved Feb. 20, 2005 from Globalspec website: <http://connectors.globalspec.com/ProductFinder/Electrical_Electronic_Components/Connectors>.

12 Fiber Optics. (2001). Retrieved 18, 2005 from Fiber Optics website: <http://www.fiber-optics.info>.

13 HARTING Homepage. (2005). Retrieved Feb. 21, 2005 from Harting website: <http://www.harting.com>.

14 Machine Design. (1995). Retrieved Feb. 18, 2005 from website: <http://www.machinedesign.com>.

# Passive

# 20 Capacitors

The basic function of a **capacitor** is to store electrical energy. A capacitor is similar to a battery in that it stores energy but is unable to produce it. Capacitors have two conducting plates separated by an insulator called the **dielectric**. The dielectric can be any type of material that keeps the two plates from touching and does not conduct electrical current.[1]

When attaching a capacitor to a battery, one plate attaches to the positive terminal and the other plate to the negative terminal. The positive terminal receives electrons from the battery and the negative terminal releases them. Once the capacitor is fully charged it has the same voltage as the voltage source in the circuit. The amount of charge the capacitor can hold is referred to as its capacitance. The capacitance is measured in units of **farad** (F) but is usually expressed in micro-farads (μf).

There are several types of capacitors that are differentiated by their dielectric material and applications. These include film, aluminum electrolytic, paraffin paper, tantalum, and ceramic. Capacitors can also be either fixed or variable. A **fixed capacitor** has a capacitance value that is nonadjustable, where as a **variable capacitor** has a capacitance value that could be changed or set by some adjustment mechanism[2].

**Film capacitors**, such as film-foil capacitors, have a plastic film dielectric wound alternately with metal foil electrodes. For example, a polycarbonate film capacitor can operate in a wide range of temperatures and is ideal for timers and filters in places of high temperature.

**Aluminum electrolytic** capacitors have a dielectric that is formed by an electrolytic process and are classified as either wet or dry. **Wet electrolytic** capacitors have a moist electrolyte and **dry or solid** capacitors do not. As well as being wet or dry, aluminum electrolyte capacitors are also referred to as leaded capacitors or chip aluminum electrolyte capacitors. **Leaded capacitors** are connected to a circuit by different kinds of leads. **Chip aluminum capacitors** do not have leads and are referred to as surface mount capacitors. They are used because of their high volumetric density as well as their ability to operate over a range of temperatures[1].

**Paraffin paper capacitors** have a thin sheet of paper, known as **kraft paper**, made from wood pulp rolled between two metal foils.

**Tantalum capacitors** have a dielectric made up tantalum. This alloy is used because of its high melting point, high strength, and good ductility. These types of capacitors are used in small applications such as pagers, cell phones, and other electronic devices.

**Ceramic capacitors** have the dielectric made of a ceramic material because of its ability to permit electrostatic attraction and repulsion to take place across it. Different ceramics are used in different applications. Ceramic COG (NPO) for instance, is used for timing and tuning applications. If a ceramic capacitor does not meet the standard for a specific operation, manufacturers often design new ones specifically for certain operations.

Recently, another dielectric material has grown in popularity: niobium. This material has similar characteristics to tantalum but is more abundant in the environment. **Niobium capacitors** are particularly popular in the notebook computer sector with growing demand from white goods (appliance) suppliers, digital camera manufacturers, cellular telephone producers, automotive telematics and entertainment, switch-mode power supply vendors, and telecom infrastructure equipment producers[5].

# The Product

## Applications and End Uses

In some cases, capacitors are used to store charge for high-speed use. For example, large lasers use this technique to produce very bright, instantaneous flashes. Capacitors can also eliminate ripples in a circuit. If a line carrying DC voltage has ripples or spikes in it, a large capacitor can stabilize the voltage by absorbing the peaks and filling in the valleys. A capacitor also has the potential to block DC voltage. "If you hook a small capacitor to a battery, then no current will flow between the poles of the battery once the capacitor charges. However, any alternating current (AC) signal flows through a capacitor unimpeded, which happens because the capacitor will charge and discharge as the alternating current fluctuates, making it appear that the alternating current is flowing."[1]

Film capacitors, with the capability to operate over a wide range of temperatures, are commonly used in cameras and timers. They are manufactured using both plastics and metals, making them difficult to recycle. Because of this, substitutes are used when possible. Aluminum capacitors are one of the most commonly manufactured and also easy to recycle. Paraffin paper capacitors are also commonly used and are cheap to manufacture. Tantalum capacitors are small capacitors with high density making them excellent for cell phones, pagers and other mobile devices where size and weight are important manufacturing points. Tantalum is a natural element but it is much rarer than aluminum, which makes it more expensive. Ceramic capacitors are very efficient in their operation and used in many timing applications[1].

## Product Features and User Benefits

**Figure 1:** Product Features and User Benefits of Capacitors

| Product Features | | User Benefits |
|---|---|---|
| Heat and Moisture Resistant | ⇨ | Able to operate at optimal level of performance in harsh environments. |
| Low Cost | ⇨ | Saves money during design as well as rework and repair. |
| Long Life | ⇨ | Low maintenance costs, greater reliability, and less downtime of the electronic device. |
| Timing Capabilities | ⇨ | Able to use in applications that require specific timing of storage and release of energy. |

## Life Cycle

**Figure 2:** Life Cycle Curve for Capacitors

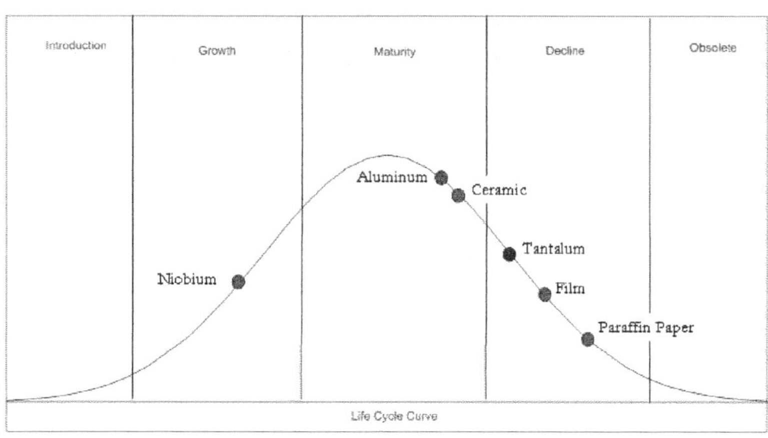

# The Market

## Market Trends

The total available market for capacitors in 2004 was $13.3 billion, up from $12.4 billion in 2003[3]. The market for capacitors has grown at a rate of 3% since 1999 and is expected to do so until 2007 when the projected TAM is expected to eclipse $14.5 billion.

**Figure 3:** Capacitor Market Trends[3]

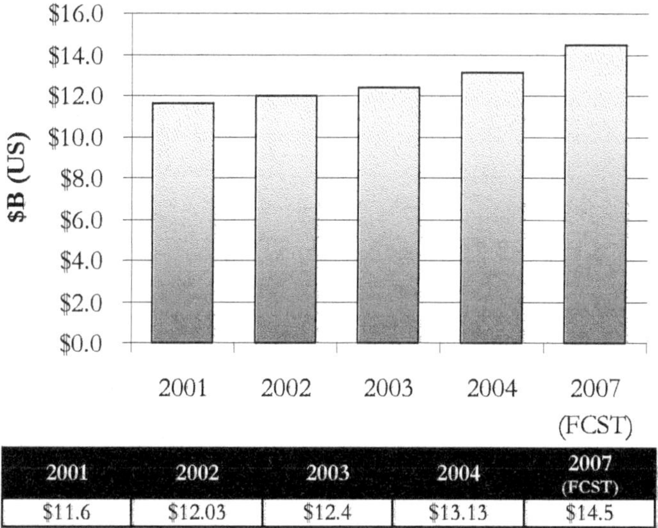


| 2001 | 2002 | 2003 | 2004 | 2007 (FCST) |
|---|---|---|---|---|
| $11.6 | $12.03 | $12.4 | $13.13 | $14.5 |

Strong competition and a declining demand for capacitors will cause a price erosion of 2% to 3% in 2005[3]. A slow but steady growth for the worldwide capacitor market is expected in 2005 with an increase in sales of tantalum capacitors expected from $2.2 billion in 2004 to $2.3 billion in 2005. Because of the extensive use of ceramic capacitors in the cell phone industry, the ceramic capacitor market is projected to grow from $3.7 billion in 2004 to nearly $3.9 billion in 2005, accounting for nearly 28% of all capacitor sales worldwide.

## Market Segmentation

**Figure 4:** Capacitor Market Segmentation for Selected Capacitors[4]

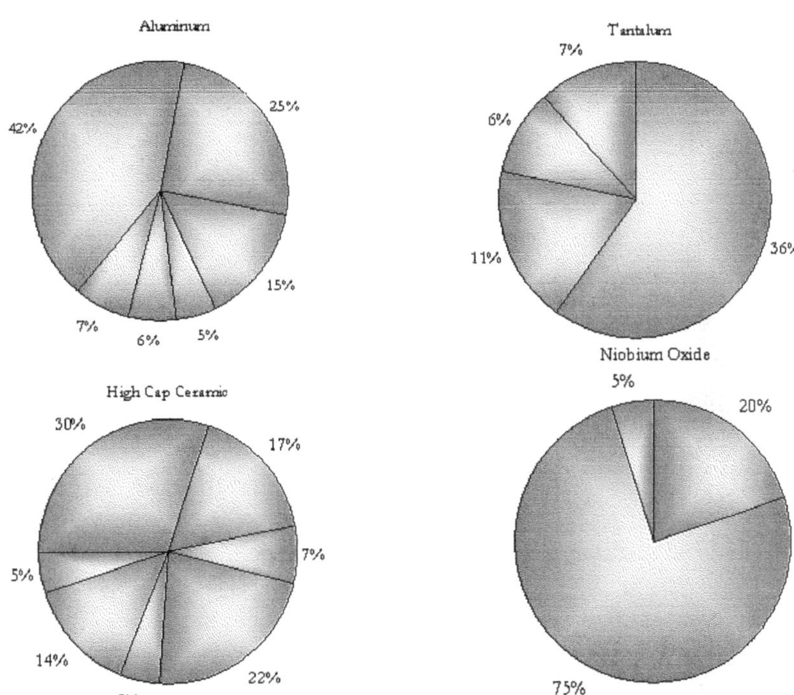


| Capacitor Type | Aluminum | Tantalum | High Cap Ceramic | Niobium Oxide |
|---|---|---|---|---|
| Consumer A/V | 42% | 18% | 30% | - |
| Wireless | - | 22% | 17% | - |
| Power | 25% | - | 7% | - |
| Computer | 15% | 36% | 22% | 20% |
| Telecom | 5% | 11% | 5% | 75% |
| Auto | 6% | 6% | 14% | 5% |
| Specialty | 7% | 7% | 5% | - |

# The Players

## Key Players

**Figure 5:** Key Players in Capacitor Market

| | Supplier | Website | Distributors | Revenue $M (US) | Market Share % |
|---|---|---|---|---|---|
| 1 | Matsushita | www.mei.co.jp | * As of June 2005, is no longer a supplier to the United States | $81,436 | 12% |
| 2 | Murata | www.murata.com | TTI | $4,015 | 8.6% |
| 3 | Nichicon Corporation | www.nichicon-us.com | TTI | $2,414 | 5.8% |
| 4 | AVX/Kyocera | www.avxcorp.com | Arrow; Avnet; Classic Components; Premier Farnell; Richardson; TTI | $1,137 | 4.2% |
| 5 | Kemet | www.kemet.com | TTI | $970 | 3.3% |
| 6 | Vishay | www.vishay.com | Arrow; Avnet; Jaco; Future; Premier Farnell; Richardson; TTI | $434 | .43% |

## References


1 Brain, M. (2005). How Capacitors Work. Retrieved Jul. 28, 2005 from How Stuff Works website: <http://electronics.howstuffworks.com/capacitor.htm>.

2 Capacitor. (2005). Retrieved Jul. 28, 2005 from Wikipedia website: <http://en.wikipedia.org/wiki/Capacitor.>.

3 Roos, G. (2004). Recovery builds for resistors, capacitors. Retrieved Jul. 28, 2005 from Purchasing website: <http://www.purchasing.com/article/CA448235.html?industryid=2147>.

4 Zogbi, D. (2005). A technical-economic analysis of the global market for capacitors between 1 and 330 mF. Retrieved Jul. 28, 2005 from TTI, Inc. website: <http://www.ttiinc.com/object/ME_Zogbi_20050606.html.>.

5 Zogbi, D. (2004). Demand for Niobium Oxide Capacitors Is Growing Rapidly. Retrieved Jul. 28, 2005 from TTI, Inc. website: <http://www.ttiinc.com/object/ME_Zogbi_20040712.html>.

# 21 Filters

A **filter** is a device that helps separate the unimportant information, commonly referred to as noise, from the desired information in real-world signals[13]. It eliminates unwanted frequencies from an electronic signal, such as random noise. **Electronic filters** are electronic devices that perform signal processing functions, including processing, amplification, and interpretation of signals. Filters will always be classified as one of the two main categories, **analog** or **digital.** However, to aid in the design process, they can be classified in any number of different ways, found in Figure 1[10].

**Figure 1:** Different Classifications of Filters[10]

| Classifications | | |
|---|---|---|
| Passive | OR | Active |
| Discrete-Time | | Continuous-Time |
| Linear | | Non-Linear |
| Infinite Impulse Response | | Finite Impulse Response |
| High Pass | | Low Pass |
| Band Pass | | Band Stop |

**Analog filters** use analog electronic circuits to produce a filtering effect on analog signals, such as electric potential, sound waves, and mechanical motion. The analog electronic circuits typically used are resistors, inductors, capacitors, and operating amplifiers[2]. Analog filters are split into two main categories: passive filters and active filters.

Digital filters perform numerical calculations on sampled values from a signal with the use of a digital processor. These filters can be constructed to mathematically control digital data to provide either the same effect as analog filters or to produce filters whose functions are foreign to the analog world[6]. The boundaries/limitations of digital filters are more stable than those of the more commonly used continuous analog filters. The two general purposes of digital filters are to separate signals that have been combined (signal separations) and to restore signals that been distorted in some fashion (signal restoration)[10]. Signal separation is needed when a transmitting signal has been contaminated with some kind of interference, noise, or other signals, whereas signal restoration is needed when a signal has been distorted in some way.

Though analog filters can be used to perform these same tasks, digital filters can achieve far superior results[7].

**Passive filters** are the simplest form of analog filters and are based on their passive component elements: resistors, inductors, and capacitors[12]. The filter is passive because it is activated by the power within the signal and not by any external power supply. The inductors within the filter prevent high-frequency signals from passing, but allow, or conduct, low-frequency signals to pass, while the capacitors perform the same operations in reverse. Low-pass and high-pass filters are both classified as passive filters. Low-pass filters operate by transmitting low-frequency signals more strongly than high-frequency signals and do this when a signal passes through an inductor, or has a path to the ground through the capacitor. High-frequency filters do the opposite but with high-frequency signals[22].

**Active filters** are a type of analog electronic filter that combines active components, usually operational amplifiers, with resistors and capacitors (passive components)[1]. Active filters more closely match ideal filters than passive filters, possessing the ability to have a higher quality, and achieve **resonance** with out the use of an inductor. Active filters have an upper-frequency threshold, limited by the **bandwidth** of the amplifiers used within the filter[1]. A **sallen-key filter** is an example of an active filter.

**Discrete-time filters** are those that do not filter continuously[8]; in contrast, **continuous time filters** are always filtering[5].

A **linear filter** will apply a linear transformation to a time-varying input signal. They are typically used to eliminate unwanted frequencies from input signals or select wanted frequencies[17]. A **non-linear filter** is a filter whose output is not a linear function of its input. Examples of non-linear filters include the Kalman filter, phase-locked loops filter, detectors, and mixers[20].

**Infinite Impulse Response Filters** or **IIR filters** are those that have an impulse response function, which is non-zero over an infinite length of time. IIR filters can be either analog or digital. Analog IIR filters are typically resistor capacitor (RC) filters, which have an exponential impulse response characterized by an RC time constant. A digital IIR filter's output feedback is immediately apparent in the equations defining the output. IIR filters tend to be quick and cheap; however, they have poor bandpass filtering and stability characteristics compared to FIR filters[16]. **Finite Impulse Response Filters** are always digital filters that are normally implemented through digital electronic computation. The **Z transform** (conversion of a discrete time into a signal) of an FIR filter has no poles, only zeros[14]. These two classifications are known as **impulse response classifications**.

The four basic **frequency responses** of filters include: high pass or low pass, and band pass or band stop. A **high pass filter** passes high frequencies well and **attenuates** or decreases the magnitude or value of low frequencies.

Therefore, a high pass filter will block unwanted lower frequencies[15]. A **low pass** filter passes low frequencies well and attenuates unwanted high frequencies[18]. A **band pass filter** passes frequencies within a desired range and attenuates frequencies not within that range. These filters can be created by combining a low pass and a high pass filter[3]. Finally, a **band stop filter** or a **notch filter** passes most frequencies unaltered but attenuates those in a narrow range to low levels. A band stop filter is the opposite of a band pass filter[21]. Each of these frequency responses can qualify as any or all of the classifications.

Other types of filters are older, created from rigid type materials. They include quartz filters, surface acoustic wave filters (SAW), garnet filters, and atomic filters[10].

**Quartz filters** were developed in the late 1930s. They have higher quality factors than basic LCR filters and are used when higher stabilities are required[10]. **Surface Acoustic Wave filters** or **SAW filters** are electromechanical filters used for radio frequency applications. The electrical signal is converted into a mechanical wave in a piezoelectric crystal; the wave propagates across the crystal and, in doing so, it is delayed or slowed down. The delayed outputs are recombined to produce a direct analog implementation of a finite impulse response filter. **Garnet filters** are used at frequencies from 800MHz to 5GHz. A synthetic garnet sphere sits on a strip of metal that is driven by a transistor, a small loop antenna touches the top of the sphere, then an electromagnet is used to change the frequency and as this happens the garnet will pass. Lastly, **atomic filters** are used for extremely high frequencies and acute precision. The vibrations of atoms are used to achieve the filtering.

# The Product

## Applications and End Uses

Analog filters are widely used in applications that require noise reduction, video signal enhancement, and graphic equalization, as well as various other areas. Commonly, analog filters are used in medical instrumentation when line frequencies need to be detected by sensors. Digital filters can perform virtually any filtering effect that can be expressed as a mathematical algorithm. Since the cost of ICs has continued to decline, digital filters have become more common and are now essential parts of objects such as radios, cell phones, and stereo receivers[7]. Digital filters are commonly used for sound quality equalization and editing for playback. They can also be used to test the durability of a signal. The **Kalman filter** is used for order tracking. Data acquisition systems use digital filters as well. Other applications include decimation, zoom transformation, interpolation, active control, communications, and differentiation[4].

## Product Features and User Benefits

**Figure 2:** Product Features and User Benefits of Analog Filters

| Product Features | | User Benefits |
|---|---|---|
| Inexpensive | ⇨ | Low unit cost allows monetary savings without sacrificing a large amount of performance. |
| Fast | ⇨ | Filters data at high rates, saving time and increasing process capabilities as a result of faster production. |
| Emphasis on Handling Limitations | ⇨ | Promotes better accuracy and stability of the resistors and capacitors to ensure more efficient filtering and better production. |
| Internal Power Supply | ⇨ | Internal power source increases ease of use, board real estate, reliability, and cost efficiency by reducing the need for external power supply. |
| Large Dynamic Range in Both Amplitude and Frequency | ⇨ | Provides flexibility as well as the ability to take on larger filtering loads. |

**Figure 3:** Product Features and User Benefits of Digital Filters

| Product Features | | User Benefits |
|---|---|---|
| Programmable | ⇨ | Pre-stored operating program in processor's memory allows the filter to be easily changed without affecting the circuitry. |
| Versatile | ⇨ | Can be easily altered to suit surrounding environments and various operating requirements. |
| Emphasis on Signal Limitations | ⇨ | Superior performance provides crisp filtering, improving the integrity of signal transmission. |
| Easily Designed, Tested, and Implemented | ⇨ | Can be designed, tested, and implemented without difficulty, saving design and test time and accelerating time to market. |
| Stable Operating Conditions | ⇨ | High reliability with flexibility to operate in a wide range of environments–temperature, noise, vibration, and so on. |

## Life Cycle

The filters displayed in the Life Cycle Curve below are a combination of the most common filters, new filters, and older filters. The common filters include Butterworth, Chebychev, and Bessel filters. The newest filter technologies are the IIR and FIR filters, including the Kalman. Lastly, the older filters include those created from rigid sources, such as garnet and quartz.

**Figure 4:** Life Cycle Curve for Selected Filter Products

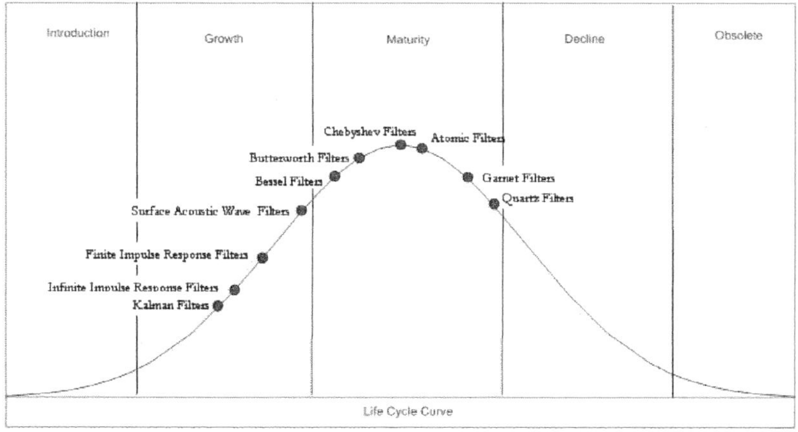

# The Market

## Market Trends

**Figure 5:** Total Available Market for Selected Filter Types[9]

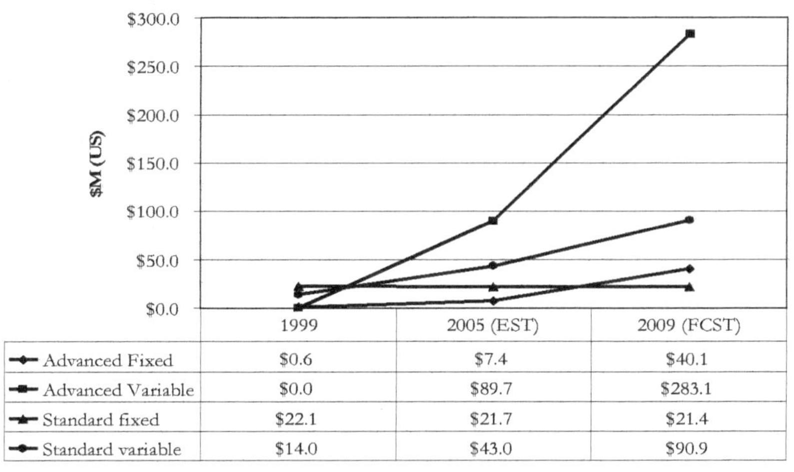


| | 1999 | 2005 (EST) | 2009 (FCST) |
|---|---|---|---|
| Advanced Fixed | $0.6 | $7.4 | $40.1 |
| Advanced Variable | $0.0 | $89.7 | $283.1 |
| Standard fixed | $22.1 | $21.7 | $21.4 |
| Standard variable | $14.0 | $43.0 | $90.9 |

## Regional Segmentation

**Figure 6:** Global production of Inductors and Filters for 2000[23]

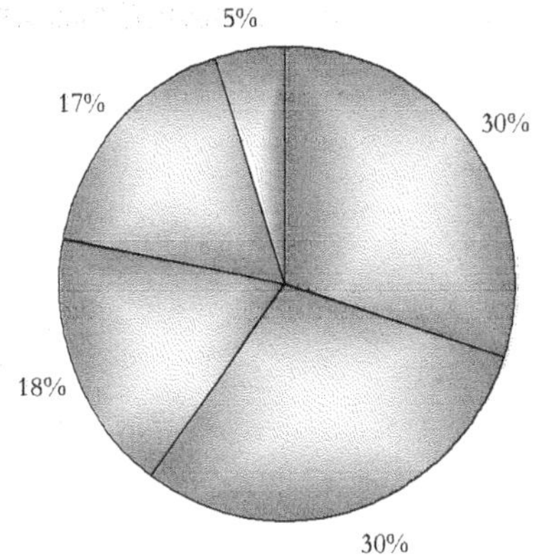


| World Production of Inductors and Filters in 2000 | |
|---|---|
| Japan | 30% |
| Other Asia | 30% |
| Europe | 18% |
| United States | 17% |
| Rest of World | 5% |

# The Players

## Key Players

**Figure 7:** Leading Suppliers of Filters for 2000

| | Supplier | Website | Distributors |
|---|---|---|---|
| 1 | Murata | www.murata-northamerica.com | Avnet; Arrow; Digi-Key; Future; Mouser; TTI |
| 2 | AVX/Kyocera | www.avxcorp.com | Allied; All Star Magnetics; Arrow; Avnet; Digi-Key; Future; Garrett; Mouser; Newark InOne; Powell; Richardson; TTI |
| 3 | TDK | www.tdk.com | All American; Arrow; Digi-Key; Future; Hammond; Jaco; MH&W; Mouser; Sager; TTI |
| 4 | Matsushita | www.matsushita.com | *As of June 2005 is no longer a supplier to the United States |
| 5 | Epcos | www.epcos.com | Adams Magnetics; Allied; Arrow: Dexter Magnetics; Digi-Key; Easter Components; Future; Newark InOne |
| 6 | Nihon Dempa Kogyo | www.ndk.com | California Capacitor; Dove; Dependable Components; Richardson |
| 7 | Nissei Arcotronics | www.arcotronics.us | Atlantic Components; Chris; First Phase; Future; Hughes-Peters; ITAL; LXI; NRC; PRidmore; Priebe; Richardson; Sager; Schuten; Trensetter; Walker |
| 8 | Toyocom | www.toyocom.com | |

## References


1 Active Filter. (2005). Retrieved Jul 27, 2005 from Wikipedia website: <http://en.wikipedia.org/wiki/Active_filter>.

2 Analog Filter. (2005). Retrieved Jul. 27, 2005 from Wikipedia website: <http://en.wikipedia.org/wiki/Analog_filter>.

3 Bandpass Filter. (2005). Retrieved Aug. 5, 2005 from Wikipedia website: <http://en.wikipedia.org/wiki/Bandpass_filter>.

4 Blough, J. (2004). Digital Filtering. Retrieved Aug. 2, 2005 from Michigan Tech University Mechanical Engineering website: <http://www.me.mtu.edu/~jrblough/5700_homepage/digitalfiltering_lecture.pdf>

5 Continuous-time. (2005). Retrieved Aug. 5, 2005 from Wikipedia website: <http://en.wikipedia.org/wiki/Continuous-time>.

6 Digital Filter. (n.d.). Retrieved Jul. 27, 2005 from Atis website: <http://www.atis.org/tg2k/_digital_filter.html>.

7 Digital Filter. (2005). Retrieved Jul. 27, 2005 from Wikipedia website: <http://en.wikipedia.org/wiki/Digital_filter>.

8 Discrete-time. (2005). Retrieved Aug. 5, 2005 from Wikipedia website: <http://en.wikipedia.org/wiki/Discrete-time>.

9 Dispersion compensators move onto optical-network fast track. (2000). *Laser Focus World WDM Solutions Supplement* (1043-8092). Penn Well Publishing.

10 Electronic Filter. (2005). Retrieved Aug. 5, 2005 from Wikipedia website: <http://en.wikipedia.org/wiki/Electronic_filter>.

11 Filter. (2005). Retrieved Jul. 19, 2005 from Answers website: <http://www.answers.com/>.

12 Filter. (n.d.). Retrieved Jul. 27, 2005 from Vision Needs website: <http://www.visionneeds.com/glossary.htm#F>.

13 Filter. (2005). Retrieved Jul. 19, 2005 from Wikipedia Web site: http://en.wikipedia.com/wiki/filter/.

14 Finite Impulse Response Filter. (2005). Retrieved Aug. 5, 2005 from Wikipedia website: <http://en.wikipedia.org/wiki/Finite_impulse_response>.

15 High-Pass Filter. (2005). Retrieved Aug. 5, 2005 from Wikipedia website: <http://en.wikipedia.org/wiki/High_pass_filter>.

16 Infinite Impulse Response Filter. (2005). Retrieved Aug. 5, 2005 from Wikipedia website: <http://en.wikipedia.org/wiki/Infinite_impulse_response>.

17 Linear Filter. (2005). Retrieved Aug. 5, 2005 from Wikipedia website: <http://en.wikipedia.org/wiki/Linear_filter>.

18 Low-Pass Filter. (2005). Retrieved Aug. 5, 2005 from Wikipedia website: <http://en.wikipedia.org/wiki/Low_pass_filter>.

19 McIlvaine, B. (2001). A different market year – Filter suppliers, customers find more. *Electronic Buyer's News.*

20 Non-Linear Filter. (2005). Retrieved Aug. 5, 2005 from Wikipedia website: <http://en.wikipedia.org/wiki/Non_linear_filter>.

21 Notch Filter. (2005). Retrieved Aug. 5, 2005 from Wikipedia website: <http://en.wikipedia.org/wiki/Band_stop_filter>.

22 Passive Component. (2005). Retrieved Jul. 27, 2005 from Wikipedia website: <http://en.wikipedia.org/wiki/Passive_filter>.

23 The Passive Component Industry in Europe. (2002). EECA-EPCIA. Retrieved Aug. 3, 2005 from EECA-EPCIA website: <http://www.eeca.org/pdf/WhiteBook.pdf>.

# 22

# Resistors

## Resistors, Potentiometers, and Trimmers

A **resistor** is a device that restricts current flow. Resistors are used in virtually all electronic circuits. Because they restrict current flow, the higher the resistance the lower the current will be after passing through the resistor. Resistance is measured in ohms, named after the renowned scientist G.S. Ohm. His theory, called Ohm's Law, states that the voltage in a circuit is equal to the current times the resistance[7]. This principle is applied by engineers when designing circuits to determine the type of resistor necessary for the circuit to function properly. Resistors can be split into two categories: variable and fixed (discrete). Resistors are also classified by the material used to manufacture them. Some of the most commonly used resistors are made from carbon film and/or metal film. When selecting a resistor, the resistance value is not the only factor that must be examined. Electric power rating and tolerance are important to consider when selecting the proper resistor. The tolerance is given in a plus/minus percentage value that determines the accuracy of the particular resistor. The **power rating** simply informs the user how much power the resistor is capable of handling. Power can be calculated by an equation similar to Ohm's Law: by multiplying the resistance times the current squared. Determining the value for power provides more information to select the resistors for the circuit.

As noted earlier, resistors are classified as either variable or fixed. The **variable resistor** is a resistor whose value can be adjusted by a mechanical movement—being turned by hand, for example, the way the volume can be changed on a radio. Variable resistors are either lower cost, single-turn types or multi-turn types with a helical element. Some even have a mechanical display to count the turns[3]. If the user does not know what resistance value is needed, then a variable resistor is helpful. Variable resistors have the ability to adjust their value easily to the user's specifications. For example, the volume knob on a radio adjusts the current flow in a variable resistor.

**Fixed resistors** are different in that their value cannot be changed. There are three main types of fixed resistors:

- Carbon Film
- Metal Film
- Wirewound

During the manufacture of **carbon film resistors**, a film layer of carbon (could be thick or thin) is deposited onto a small ceramic rod. The resistive coating is spiraled away in an automatic machine until the resistance between the two ends of the rod is as close as possible to the correct value. Metal leads and end caps are added, the resistor is covered with an insulating coating, and finally painted with colored bands to indicate the resistor value. Carbon film resistors are cheap and readily available, with values within ±10% or ±5% of their marked or "nominal" value. **Metal film** and **metal oxide resistors** are manufactured in a similar way, but can be specified more accurately to within ±2% or ±1% of their nominal value[4]. A **wirewound** resistor is made of metal resistance wire, enabling higher precision devices. Wirewound resistors cannot be used for high-frequency circuits and can become extremely hot when used for high power applications. This must be taken into consideration when designing the circuit. Ceramic resistors, a specific type of wirewound resistor, have extremely high power ratings. Ceramic resistors are simply wirewound resistors in a ceramic case and strengthened with special cement[8].

A **potentiometer**, or a pot, is simply an adjustable resistor that is used to control current flow. A potentiometer is sometimes called a **voltage divider**. Several precision control devices are considered potentiometers, but there are only two main mechanical types of potentiometers: rotary and linear[5].

Rotary pots are the most common in the potentiometer family. A **rotary potentiometer** is a rotating knob or shaft used to control the amount of current that flows through a circuit[5]. This should sound familiar considering almost every radio, stove, and dishwasher uses a rotary pot as a control device. There are two major types of rotary pots: single-turn and multi-turn. **Single-turn potentiometers** are used as angle position sensors or as panel controllers when the rotation is less than a full revolution[4]. A **multi-turn potentiometer** is used when a higher accuracy in adjustment, control, and measuring is desired or when a measurement application requires more than one revolution[4].

**Linear potentiometers** are different from rotary pots in that instead of rotating they extend and retract to control the flow of current[4]. Also included in this category are **slide potentiometers**, which serve the same purpose but are adjusted by sliding a knob up and down a rail. These types of pots can be found on some air conditioning control units.

A **digital potentiometer**, or a **digipot**, has the same functionality that a mechanical potentiometer provides but with more performance capabilities. Where the mechanical pot uses a wiper, the digital pot uses logical inputs. The digital potentiometer has a few advantages over the mechanical version, such as remote process control, remote calibration, analog to digital conversions, digital to analog conversions, variable gain amplification, and variable oscillation. A good example of a digital potentiometer is the volume control on new stereo systems. While a digital pot provides no difference in function over the mechanical version, it enables remote control.

A **trimming potentiometer**, or **trimmer**, is very similar to a rotary pot, but is used in trimming the resistance in electronic circuits and is usually mounted on a Printed Circuit Board (PCB)[1]. Instead of rotating, a trimmer, like a rotary pot, is turned. It must be turned with a screwdriver or other appropriate tool. Similar to rotary pots, trimmers are offered primarily with single-turn and multi-turn options. The single-turn trimmer provides less accuracy than a multi-turn due to the multi-turn's precision. The multi-turn also serves as a dual trimmer, making it more efficient than the single-turn.

A resistor's value is determined by a color code that enables the user to identify the characteristics of the part easily and simplifies understanding of the nomenclature. Each different colored band on the resistor represents a certain number. The table in Figure 1 shows each number—from zero to nine—being assigned a specific color. The first band from the left on the resistor represents the first digit of the resistor value. For instance, if the color of the first band is green, it represents 5. The second band from the left represents the second digit. For instance if the color is blue, it represents the number 6. In some cases, the resistor can have more than two numbers representing its value. Usually the third band from the left does not represent a number (unless in the situation of the previous sentence), but is known as the multiplier. For instance, if the third band color is orange then you will multiply the value determined by the first two bands by ten[6]. The final band represents the tolerance of the resistor. The tolerance is the percentage accuracy of the resistor's value. In the case of the tolerance values, the colors represent different values than the previous bands. A brown band indicates a ±1% tolerance, red indicates a ±2% tolerance, gold indicates a ±5% tolerance, and silver indicates a ±10% tolerance. The most common level of tolerance is ±5% indicated by a gold band.

**Figure 1**: Resistor with Color Code Table[6]

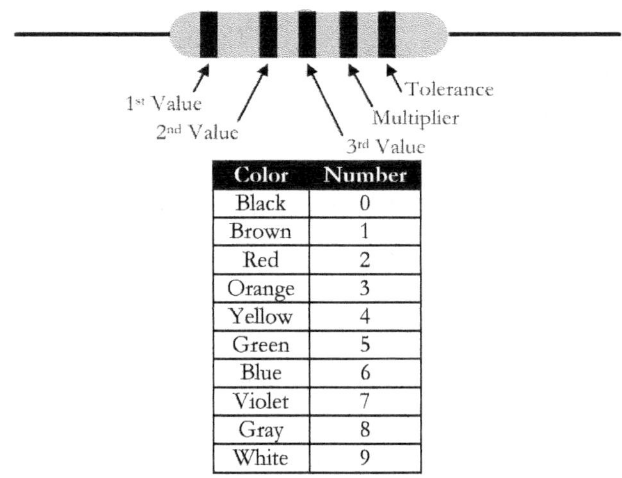


| Color | Number |
|---|---|
| Black | 0 |
| Brown | 1 |
| Red | 2 |
| Orange | 3 |
| Yellow | 4 |
| Green | 5 |
| Blue | 6 |
| Violet | 7 |
| Gray | 8 |
| White | 9 |

Resistors are placed in either series or parallel within a circuit. When resistors are placed in **series**, they have current flowing through the resistors one after the other. Since there is only one path for the current to follow in this configuration, the value of the current is the same across each resistor. Figure 2 shows a representation of a series circuit. The voltage drops across the resistors must add up to the total voltage supplied by the battery[2]. In order for the circuit to function properly, Ohms law must be satisfied.

**Figure 2:** Resistors in Series Circuit[2]

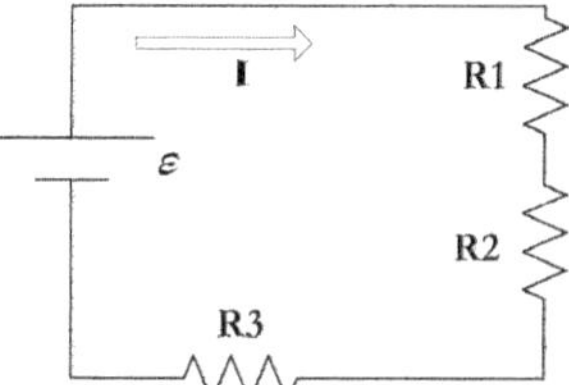


Resistors can be connected such that they branch out from a single point, known as a **node**, and join up again somewhere in the circuit. This configuration is called a parallel connection[2]. If the current has more than one path to follow, it is considered a parallel circuit. The two circuits below are good examples of parallel circuits. In parallel circuits, the total resistance is not simply the sum of the voltage drops across each resistor. To find total resistance the following equation must be applied: 1/(1/R1+1/R2+1/Rn).

**Figure 3:** Resistors in Parallel Circuit[2]

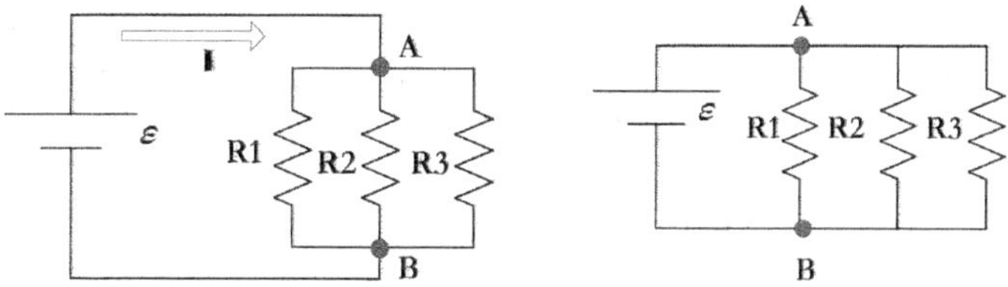


# The Product

## Applications and End Uses

Resistors are used in a number of the devices and equipment used every day. Some of the end-equipment applications of resistors are: automotive, test and measurement equipment, medical equipment, industrial equipment, and consumer electronics. In the automotive industry, resistors are used in computer chips that are in the engine. They are also used in devices such as stereo or navigation systems that are present in some vehicles. Vehicles might have a resistor in an electronic control panel for the lights, windows, and lock operations, as well as in a keyless entry remote. Test and measurement equipment include products such as amp meters and voltage testers. Medical equipment includes items such as pacemakers, cardiographs, and computers that are used to monitor patients in hospitals. Industrial equipment use includes resistors in most all types of machinery, such as a handheld remote that operates a crane. Consumer electronics uses resistors in everyday items such as cell phones, MP3 players, computers, DVD players, and palm pilots[9].

Rotary potentiometers are turned to control the resistance level of the circuits in a device. Some common uses are to control guitar amps, stereo applications, car radios, and other consumer goods that have adjustable levels of performance. Linear potentiometers are commonly used in stereo equipment such as mixers and in the high-speed automotive sector.

Digital potentiometers are the latest type and can be found in a range of new electronic devices. In particular, they can be controlled with a remote, have a much higher resolution, and can be fine-tuned to the user's specifications. Trimming potentiometers are commonly used in knobs and have a shaft that must be inserted to be turned. Trimmers are often mounted on Printed Circuit Boards (PCBs) and used to control resistance in the circuits of an electronic device. A trimming potentiometer can be used in power supplies to adjust the output voltage or current limit, in amplifiers to adjust gain control, in digital circuits as a time delay device, to adjust for errors in photocell circuits, and in instrumentation to calibrate digital voltmeters.

## Product Features and User Benefits

**Figure 4:** Product Features and User Benefits of Resistors[3]

| Product Features | | User Benefits |
|---|---|---|
| Small Size | ⇨ | Decreased board space resulting in decreased overall size of the product. |
| Low Cost | ⇨ | Saves money during design as well as rework and repair. |
| Heat Dissipation | ⇨ | Prevents circuit from overheating and causing potential system damage. |
| Moisture Resistant | ⇨ | Able to operate effectively in harsh, humid environments. |
| Maximum Current Rating Provided | ⇨ | A resistor has the maximum current rating on the component (resistor)–facilitates engineering specification activities. |

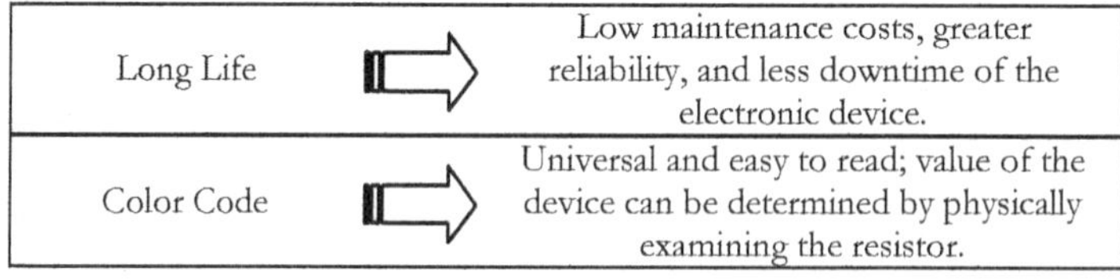

| | |
|---|---|
| Long Life | Low maintenance costs, greater reliability, and less downtime of the electronic device. |
| Color Code | Universal and easy to read; value of the device can be determined by physically examining the resistor. |

**Figure 5:** Product Features and User Benefits of Potentiometers

| Product Features | User Benefits |
|---|---|
| Can be Made Sealed or Unsealed | For use in harsh environments, a sealed pot will keep components undamaged. |
| Different Kinds of Resistive Elements | Can be made from cermet, conductive plastic, carbon, wirewound, and hybrid. This allows the pot to be designed specifically for the application. |
| Long Rotational Life | Pots can be used in applications for long periods of time without replacement. |
| Wiper Position Memory | The wiper on digipots can be stored in memory for cases where power is removed. Pot does not need to be reset in case of power outage. |
| Gold Plated Terminals | Allows for maximum capacity, which creates a better performing pot. |
| Three Mounting Types for Terminals | Enables the user to select the best pot for the application. |
| Custom Packaging | Pots can be packaged according the customer's specifications. |

## Life Cycle

**Figure 6:** Life Cycle Curve for Resistors

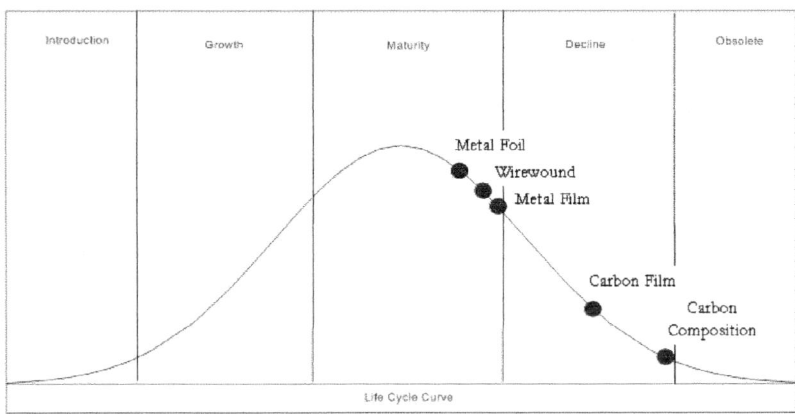


**Figure 7:** Product Life Cycle for Potentiometers

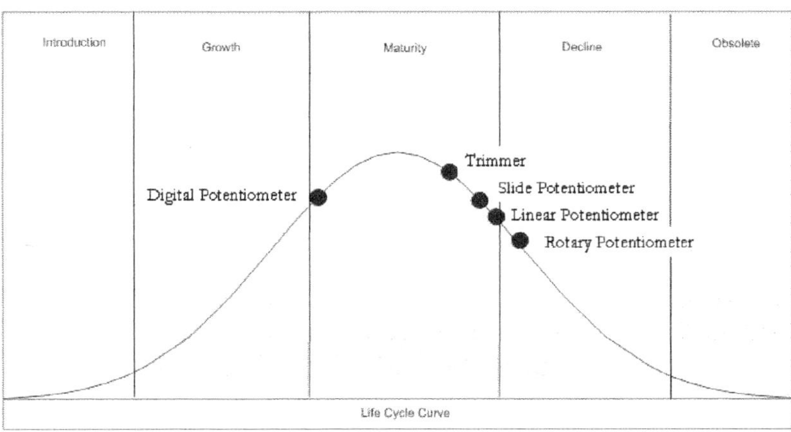

# The Market

## Market Trends

The market has, for the most part, been historically stable. In late 2001 and early 2002, resistors—like other electronics components—experienced a market decline. Since 2002, the market has slowly recovered and is predicted to continue to grow over the next three to four years. Growth of the resistor market will continue from 2004 to 2007. However, it will be hindered due to unit price declines rather than market demand decreases. Figure 7 represents the resistor component market in 2001 and 2004 and the forecast for 2007 and 2010[10].

**Figure 8:** Total Available Market for Resistor Components[10]

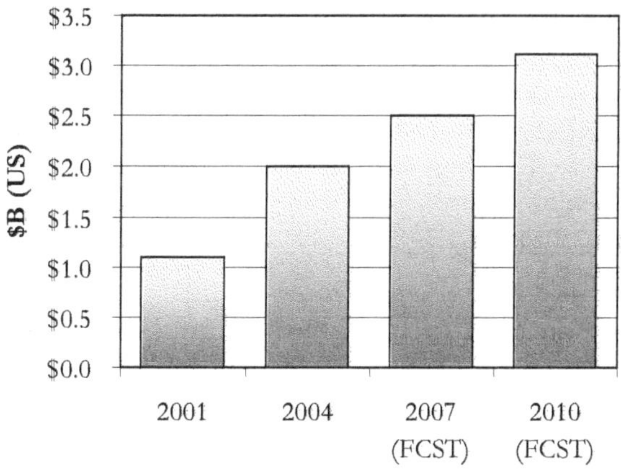


| 2001 | 2004 | 2007 (FCST) | 2010 (FCST) |
|---|---|---|---|
| $1.10 | $2.0 | $2.50 | $3.13 |

# The Players

## Key Players

**Figure 9:** Key Players in Resistor Market for 2004

| | Supplier | Website | Distributors | Revenue $M (US) |
|---|---|---|---|---|
| 1 | Yageo/Phycomp | www.yageo.com | All American; TTI | $16,710 |
| 2 | TDK | www.tdk.com | All American; Classic Components; Jaco; TTI | $6,234 |
| 3 | Vishay | www.vishay.com | Arrow; Avnet; Jaco; Premier Farnell; Future; Richardson; TTI | $2,414 |
| 4 | AVX/Kyocera | www.avxcorp.com | Arrow; Avnet; Classic Components; Premier Farnell; Richardson; TTI | $1,137 |
| 5 | KOA Corporation | www.koanet.co.jp | Classic Components; Jaco; TTI | $454 |
| 6 | Bourns | www.bourns.com | Allied; Arrow; Avnet; Premier Farnell; TTI | $387 |

## References

1 Brief Definition of Variable Resistor Products. (n.d.). Retrieved Aug. 1, 2005 from the Variable Electronic Components Institute website: <http://www.veci-vrci.com/potdefs.htm>.

2 Department of Physics. (n.d.). Retrieved Jul. 22, 2005 from University of Guelph website: <http://www.physics.uoguelph.ca>.

3 Fixed Resistors. (n.d.). Retrieved Jul. 22, 2005 from Piotech website: <http://www.piotech.com.tw/resistor.htm>.

4 P3 America: Precision Positioning Products. (2004). Retrieved Aug. 1, 2005 from P3 America website: <http://www.p3america.com/>.

5 Potentiometer. (2005). Retrieved Jul. 29, 2005 from Answers website: <http://www.answers.com/topic/potentiometer?hl=rotary&hl=potentiometer>.

6 Resistor. (2005). Retrieved Jul. 22, 2005 from Wikipedia website: <http://en.wikipedia.com/wiki/Resistor.html>.

7 Resistors. (n.d.) Retrieved Jul. 22, 2005 from Doctronics website: <http://www.doctronics.co.uk/resistor.htm>.

8 Resistors. (1998). Retrieved Jul. 22, 2005 from the Hobby of Electronic Circuit Engineering website: <http://www.interq.or.jp/japan/se-inoue/e_resistor.htm>.

9 Resistors. (2005). Retrieved Jul. 22, 2005 from TTI, Inc. website: <http://www.ttiinc.com/page/Products_Resistors.html>.

10 Weaker demand means slower resistor growth. (2005). Retrieved Jul. 22, 2005 from Purchasing website: <http://www.purchasing.com/article/CA602155.html>.

# 23 Sensors and Encoders

The basic idea of the sensor was discovered in 1947 with the bipolar transistor. Over the next 13 years, the sensor continued in its discovery phase with the Piezoresistive effect in silicon and germanium chemical elements. Finally, in 1958, the first industrial applications of the Piezoresistive effect were made. From 1960 to 1970, the strain gauge was developed and used directly on silicon. The years from 1970 to 1980 marked the batch process phase of the sensor, with batch processing of the silicon strain gauges and the first high-production sensor. Finally, 1980 to present is considered the micromachining phase: Sensors have micron-sized features, complex structures, and integration with electronics[1].

The encoder was created in the mid-1980s for the purpose of replacing the potentiometer. The substitution was needed to move to digital from analog: to the encoder's digital signal rather than the analog output created by the potentiometer[31]. Encoders serve an extensive range of functions with a large depth of products. Some standardization among encoders has been set by The Variable Electronic Components Institute, which has created and is still establishing test benchmarks for encoders[25].

A **sensor** is a device that measures a physical quantity such as heat, light, sound, pressure, magnetism, or motion[33] and then converts that signal into an electrical signal[11]. Sensors collect and measure data regarding some property of an object or material—for example, detecting the movement of a box on a conveyor. A sensor is almost an extension of humans' natural senses: sight, sound, touch, taste, and smell[5]. Terms for sensors, most of which are used interchangeably[30], include: transducer, meter, detector, and gage.

Sensors are categorized a number of different ways, one being analog and digital[32]. An **analog sensor** is actually wired into a circuit so that its output falls within a certain range. It provides a signal that is continuous in both its amount and time or space content[8]. If the output signal a sensor provides is continuous and directly proportional to the input signal, it is an analog sensor[8]. A **digital sensor**, on the other hand, sends a "discrete" signal, which means that any range of values can be the output but, they must increase in steps[3]. A digital sensor is most simply a binary device, meaning it only goes on or off, and it exists at only discrete values of time or space[8].

A second way to classify sensors is by the material being sensed. There are three generic classifications of sensors and they each can be divided into even more specific arrangements. Typical signals that are received include mechanical signals such as fluids and friction, thermal signals, electrical signals, magnetic signals, radiant signals, and chemical signals.

For market purposes, sensors are based on four different kinds: active, passive, electromechanical, and semiconductive (MEMS). The separation between active and passive, specifically, is based on the need of external active circuitry to produce the electrical output signal from the sensor. Conventionally, in the industry, it is most common to classify the sensor based on the external circuit requirement.

**Figure 1:** Four Types of Sensors[25]

| Signal Category | Definition |
|---|---|
| Active Sensors | A measurement device that adds energy to the environment during its measurement process[14], for example, "radar." A pulse is sent out, it hits an object, and it returns with a reflected signal from the original pulse[10]. |
| Passive Sensors | A measuring device that removes energy from the environment. It does not add energy as part of its measuring process[14], for example, an eyeball. Light is reflected off the surface of an object and into the sensor (eye), allowing the eye(s) to see objects[10]. |
| Electromechanical Sensors | Sensors that sense based on an entity changing from (1) an "at-rest" position to (2) an activating position that causes an event to happen, facilitating the closing of a switch either mechanically or magnetically[26]. |
| Semiconductor/MEMS Sensors | Sensors that use semiconductor technology[25]. |

Sensors are not typically found alone—usually they are operating as a part of a system that might include a signal conditioner and other analog and digital signal processing circuits—for example, a measurement system, a data acquisition system, or a process control system[18].

An **encoder** is a device that coverts linear or rotary movement or positioning into a digital signal[4]. Rick Korte, president of Heidenhain Corporation says it best: "Machines that move need a means of measuring movement" (Korte, R). Equipment has evolved to become highly complex and highly automated. Their requirement for internal measuring is mostly commonly met using the encoder.

The four basic types of encoders are:

- Rotary
- Linear
- Optical
- Magnetic

The rotary and the linear encoder are very similar in principle; their applications, however, are not. Modern linear and rotary encoders operate on the idea of the photoelectrical scanning of very fine gratings; they typically use an optical sensor that gives pulses that are translated into movement, location, or direction. **Rotary encoders** are designed to convert angular positions and motions into electrical signals; however, they are also used for linear movements. Figure 3 describes the two basic types of rotary encoders and their characteristics[20].

**Figure 2:** Incremental and Absolute Rotary Encoders[20]

| Incremental Rotary Encoders | Absolute Rotary Encoders |
|---|---|
| Rotary encoders for measuring angles | Single-turn rotary encoders (used to measure only one rotation). |
| Rotary encoders for measuring linear position | Multi-turn rotary encoders (used to measure more than one rotation). |

**Incremental rotary encoders** typically have two pulses that are out of phase with each other, allowing for the measurement of not only motion but also rotation. A third pulse, the command pulse, counts the number of revolutions in a rotating device[19].

**Absolute rotary encoders** are designed to retain its application's position after power has been removed and then restored again. This type of encoder has opaque and transparent segments that are alternating and it uses multiple groups of segments that form concentric circles on the encoder wheel. The advantage to absolute encoders is that they can be geared down so that the encoder wheel makes one revolution during the full length of machine travel. This means that the encoder does not need to be calibrated each time power is removed—it knows the exact position of the machine[19].

**Linear encoders** are used to measure linear movements. This technology is so sophisticated that the results are in the "submicron" range, meaning they are incredibly precise. Linear encoders are similar to their sister, the rotary encoder, because they operate on the same principle. Unlike rotary encoders, the linear encoder's construction is straight and the output signals are interpolated or digitized in a direct manner[20]. There are two versions of a linear scale: exposed and enclosed. Figure 4 gives an explanation of each as well as example applications.

**Figure 3:** Exposed and Enclosed Linear Encoders[20]

| Enclosed Linear Encoders | Exposed Linear Encoders |
|---|---|
| The scanning unit and glass scale are together; both are protected by "sealing lips" from contamination. | The scanning unit and glass scale are physically separated, providing for easier mounting and higher traversing speeds (no friction). |
| Applications include: machine tools, cutting-type machines, and any machine located in an extreme environment. | Applications include: coordinate measuring machines, translation stages, and material handling equipment. |

When selecting a linear encoder, one must not only consider length and dimensional profile. Permissible traversing speed, accuracy and resolution requirements, thermal behavior, and mounting guidelines must also be taken into account.

Linear and rotary encoders alike are increasingly being used as sensors for gathering angular and linear positioning as well as for collecting data in travel, measurement and control, machine tools, industrial robots, and many other sectors of technology[20].

**Optical encoders** are used to provide a signal output for the positioning information of a rotating wheel[12]. They consist of a rotating disk, a light source, and a photodetector. The process of an optical encoder begins with the disk rotating; the coded patterns of opaque and transparent sectors on the disk interrupt the light emitted onto the photodetector, generating a digital or pulse signal output[24].

A **magnetic encoder** has a rotating gear made of a ferrous metal and a magnetic pick-up that contains a permanent magnet and a sensing element. To generate a digital or pulse signal output, the gear has precisely machined teeth that provide a code pattern. As the gear rotates, the teeth interrupt the magnetic flux emitted by the permanent magnet, causing the flux to expand and collapse. The two basic kinds of magnetic pick-ups are **hall effect** and **variable reluctance**, shown in Figure 5[21].

**Figure 4:** Magnetic Pick-Ups[21]

| Magnetic Pick-up | Definition |
|---|---|
| Hall Effect | Generates a voltage by applying an external magnetic field perpendicularly to the direction of the current in a semiconductor[2]. |
| Variable Reluctance | Uses a coil of wire in the magnetic field. As teeth pass by they change the reluctance of the system, inducing a voltage pulse in the coil that is proportional to the rate flux change. |

# The Product

## Application and End Uses

The major industries sensors are used in include the automotive, consumer, medical, and safety/security[15] industries.

In the automotive industry, sensors are used in end products such as the speedometer, which measures how fast a car is going and its rate of acceleration[15]. One of the newer applications is in car seats to detect who the passenger is and determine if an airbag should be deployed or not. A car's air conditioner is equipped with sensors to measure the in-car temperature[15]. Consumers use sensors in their exercise equipment and even in children's toys. Digital cameras, cell phones, MP3 players, and other consumer electronics also use sensors[9].

Medical sensors are used for MRI and CAT scanning machinery. Other medical applications include heart-rate monitors, blood diagnostics, blood pressure cuffs, urine analysis, and dental color matching[15]. Security/safety applications include fingerprint or retina identification sensors, motion and tracking sensors for vehicle and human identification, and door and window sensors for intrusion-prevention purposes[28].

Those listed in Figure 6 are some of the most commonly found and used. For other sensors, visit a sensor manufacturer's website to view its product list.

**Figure 5:** Different Types of Sensors by Object Sensed[27]

| Sensor | Examples |
|---|---|
| Light Sensor | Photocells, photodiodes, photo-electric tubes, image sensors |
| Sound Sensors | Microphones, hydrophones, seismic sensors |
| Temperature Sensors | Thermometers, thermocouples, thermistors |
| Heat Sensors | Bolometer, calorimeter |
| Radiation Sensors | Geiger counter, dosimeter |
| Subatomic Particle Sensors | Scintillometer, cloud chamber, bubble chamber |
| Electrical Resistance Sensors | Ohmmeter, multimeter |
| Electrical Current Sensors | Galvanometer, ammeter |
| Electrical Voltage Sensors | Leaf electroscope, voltmeter |
| Electrical Power Sensors | Watt hour meters |
| Magnetism Sensors | Magnetic compass, hall effect device |
| Pressure Sensors | Barometer, pressure gauge, air speed indicator |

| Gas and Liquid Flow Sensors | Anemometer, flow meter, gas meter, water meter |
| --- | --- |
| Chemical Sensors | pH glass electrodes, lambda sensors, ion-selective electrodes |
| Motion Sensors | Radar gun, speedometer, tachometer, odometer |
| Orientation Sensors | Gyroscope, artificial horizon |
| Mechanical Sensors | Position sensor, switch, strain gauge |

The major applications of encoders include being mounted to positioning tables, screw drives, gearheads, machining tools, or almost any other application where a rotary actuator exists. The range of rotary encoders is so broad that a different encoder exists for almost every specific application that might require position response[13].

Encoders are also used in scanning medical instruments for MRI, x-ray, and nuclear imaging systems, in addition to fluid analysis devices. They are used in radio and optical telescopes, radar pedestals, tracking mounts, and communication antennas for tracking and positioning in the sky. Finally, encoders are used in high-performance printing and graphic arts industries[13].

## Product Features and User Benefits

Many different types of sensors are used in various industries all over the world and each sensor will have its own specific features and user benefits. However, some specifications span across all sensors, the most important traits being sensitivity, stability, and repeatability[30].

**Figure 6:** Product Features and User Benefits of Sensors[30]

| Product Features | User Benefits |
| --- | --- |
| Sensitivity | A highly sensitive device only measures what it is designed to measure. Temperature range and cost are two examples. |
| Stability | A stable device is one that remains in the same form under expected and reasonable conditions of storage or use[7]. Examples are corrosion and life length. |
| Repeatability | A repeatable device is one that has the ability to reproduce its outputs when the same inputs are applied to it consecutively[17]. |

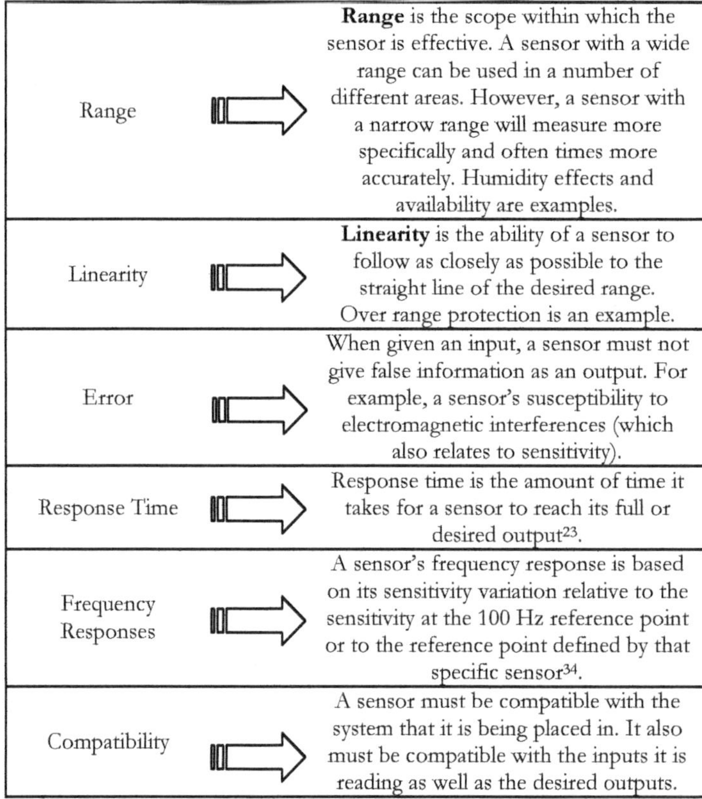

| | |
|---|---|
| Range | **Range** is the scope within which the sensor is effective. A sensor with a wide range can be used in a number of different areas. However, a sensor with a narrow range will measure more specifically and often times more accurately. Humidity effects and availability are examples. |
| Linearity | **Linearity** is the ability of a sensor to follow as closely as possible to the straight line of the desired range. Over range protection is an example. |
| Error | When given an input, a sensor must not give false information as an output. For example, a sensor's susceptibility to electromagnetic interferences (which also relates to sensitivity). |
| Response Time | Response time is the amount of time it takes for a sensor to reach its full or desired output[23]. |
| Frequency Responses | A sensor's frequency response is based on its sensitivity variation relative to the sensitivity at the 100 Hz reference point or to the reference point defined by that specific sensor[34]. |
| Compatibility | A sensor must be compatible with the system that it is being placed in. It also must be compatible with the inputs it is reading as well as the desired outputs. |

**Figure 7**: Product Features and User Benefits of Encoders[6]

| Product Features | User Benefits |
|---|---|
| Reverse Polarity Protection | Allows the encoder to be connected to the power source with regard to the polarity of the power source's connection. |
| Read in Absolute Gray Code, Natural Binary, or Binary Coded Decimal | Allows the user to use several different coding systems when converting the input data. |
| Versatile Direction Control | Can be run in both clockwise and counter-clockwise and can be changed to fit the application. |
| Operating Speed | Have a wide range of operating speed that can range from as low as 30 RPM to over 1000RPM to fit the desired application. |

| | | |
|---|---|---|
| Field Programmable Zero Position | ⇨ | Can be set to start reading at any position in its rotation and the start (zero) position can be adjusted to the desired location. |
| 360 Degree Visible Range | ⇨ | Capable of reading data at any position in its rotation, giving it a full range of visibility. |
| High Degree of Accuracy | ⇨ | Extremely accurate capabilities that allow for very few errors to take place. |
| Long Lasting | ⇨ | Have a long life expectancy and can save money for the user in the long run. |

## Product Trends

The new trends in the sensor industry are smart sensors, soft sensors, and micro-electromechanical sensors, or MEMS. **Smart sensors**, or **intelligent sensors**, are sensors that contain a signal processor and intelligence capabilities in a single chip. These can be seen as pressure sensors and accelerometers, biosensors, chemical sensors, optical sensors, and magnetic sensors. **Soft sensors** are those that do not have any sensing hardware as a part of the product. They use a microprocessor that processes data acquired from different devices and then combines it using a mathematical model, producing an estimation of the sensing interest. **MEMS** are currently one of the most active sensor research fields; it combines micromachining technology with semiconductor processing technology (mostly IC technology) that provides for a number of different sensors.

The trends in the encoder industry include fieldbus encoders, programmable smart encoders, and hollow-shaft encoders. A **fieldbus encoder** is designed to use distributed control, which includes a common interconnection to transport the data signals for the entire motion control system. They have extensive installation time and cable costs, but the components themselves can be implemented in a "plug-and-play" manner that will reduce technical support costs.

**Programmable smart encoders** are encoders with an extra internal processor that makes calculations based on the initial value provided by the optical or magnetic based system. They are versatile: One programmable smart encoder can be used in several different applications. Many parameters can be programmed into the encoder, including resolution, output code, and number of turns—even a fractional number. All encoders are coupled in two places: the shaft and the body of the encoder. **Hollow-shaft encoders** are stator coupled, which differs from the shaft-style encoders that are rotor coupled. Stator coupling provides the mechanical coupling between the encoder and the motor on the "housing side" of the encoder, therefore the stator side will only see the very low friction in the bearings[22].

## Life Cycle

The life cycle of a sensor is difficult to estimate due to the hundreds of different types, each with its own individual life cycle. Figure 9 depicts the Life Cycle Curve for a few of the technologies of sensors. Those found in the introduction and growth stages of the product life cycle are all new trends in the industry; the bipolar transistor and strain gauge were the first sensors.

**Figure 8:** Product Life Cycle Curve for Sensors by Technology/Classification

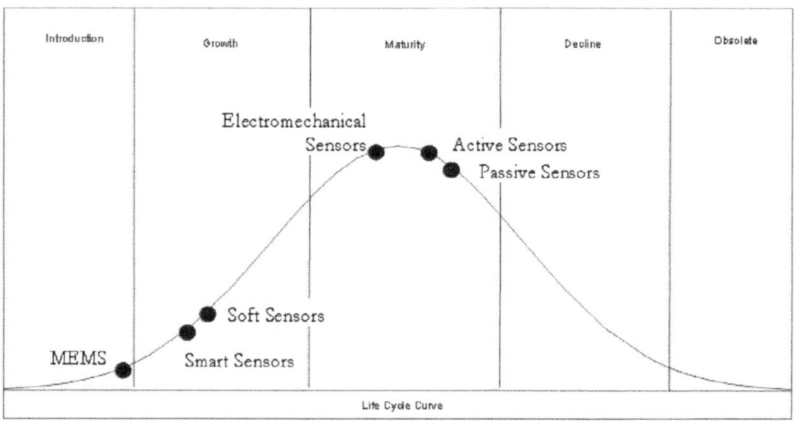


**Figure 9:** Product Life Cycle for Encoders

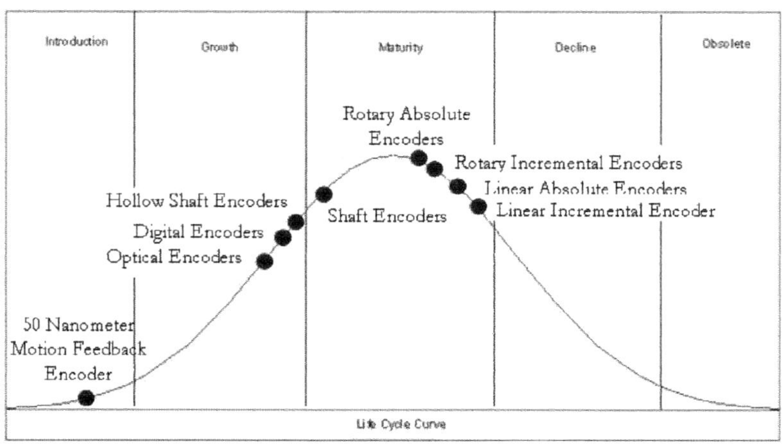

# The Market

## Market Trends

**Figure 10:** Total Available Market for Sensors[25]

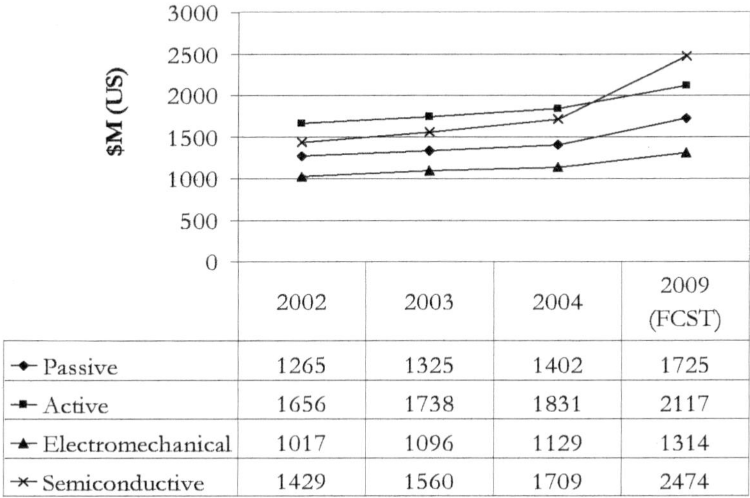


| | 2002 | 2003 | 2004 | 2009 (FCST) |
|---|---|---|---|---|
| Passive | 1265 | 1325 | 1402 | 1725 |
| Active | 1656 | 1738 | 1831 | 2117 |
| Electromechanical | 1017 | 1096 | 1129 | 1314 |
| Semiconductive | 1429 | 1560 | 1709 | 2474 |

## Market Segmentation

**Figure 11:** Market Segmentation for Sensors and Actuators

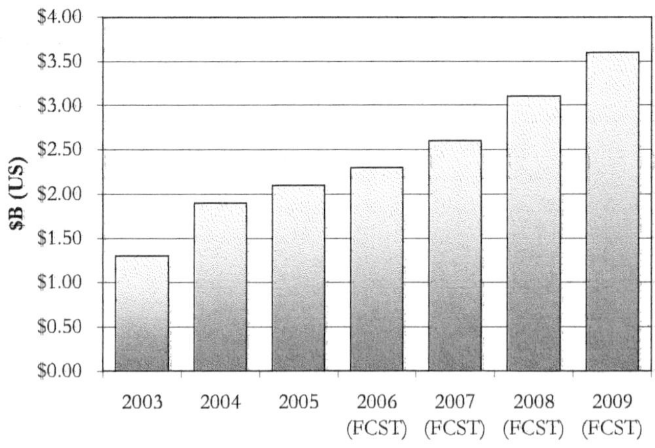


Source: iSuppli Corporation

| 2003 | 2004 | 2005 | 2006 (FCST) | 2007 (FCST) | 2008 (FCST) | 2009 (FCST) |
|---|---|---|---|---|---|---|
| $1.30 | $1.90 | $2.10 | $2.30 | $2.60 | $3.10 | $3.60 |

# The Players

## Key Players

**Figure 12:** List of the Major Manufacturers of Sensors

| Company | Headquarters |
|---|---|
| Agilent Technologies Inc. / Semiconductor Products Response Center (Manuf. & Service) | Santa Clara, CA |
| All Sensors (Manufacturer) | San Jose, CA |
| Allegro MicroSystems, Inc. (Manufacturer) | Worcester, MA |
| Analog Devices (Manuf. & Service) | Norwood, MA |
| Andigilog (Manufacturer) | Tempe, AZ |
| austriamicrosystems USA, Inc. (Manufacturer) | San Jose, CA |
| Beacon Dynamics (Manufacturer) | Dover, NJ |
| Bill West, Inc. (Manuf., Distrib., Sole Distrib. & Service) | Monroe, CT |
| Building Automation Products, Inc. (Manufacturer) | Gays Mills, WI |
| CeramTec North America (Manuf. & Service) | Laurens, SC |
| Colibrys (Manuf. & Service) | Switzerland |
| Dataprobe, Inc. (Manufacturer) | Paramus, NJ |
| Fraunhofer-IMS (Manuf. & Service) | Germany |
| Fred V. Fowler Co., Inc. (Manuf., Distrib. & Sole Distrib.) | Newton, MA |
| Freescale Semiconductor, Inc. (Manuf. & Service) | Schaumburg, IL |
| Fujitsu Microelectronics America, Inc. (Manufacturer) | Sunnyvale, CA |
| HDK America Inc. (Manufacturer) | Barrington, IL |
| HELP Temperatura & Metrologia (Manufacturer) | Brazil |
| Infineon Technologies Corporation (Manufacturer) | San Jose, CA |
| Lake Shore Cryotronics, Inc. (Manufacturer) | Westerville, OH |
| LSI Computer Systems, Inc. (Manufacturer) | Melville, NY |
| Maxim Integrated Products, Inc. (Manufacturer) | Sunnyvale, CA |
| Melexis, Inc. (Manuf. & Service) | Concord, NH |
| Merit Sensor Systems (Manuf. & Service) | Santa Clara, CA |
| Micro Commercial Components (Manufacturer) | Chatsworth, CA |
| MicroSensors, Inc. (Manuf. & Service) | Costa Mesa, CA |

| | |
|---|---|
| Mitsubishi Electric & Electronics USA, Inc. (Manufacturer) | Sunnyvale, CA |
| NanoAmp Solutions, Inc. (Manufacturer) | Milpitas, CA |
| NVE Corporation (Manufacturer) | Eden Prairie, MN |
| Omron Electronics LLC (Manuf. & Service) | Schaumburg, IL |
| Optek Technology, Inc. (Manufacturer) | Carrollton, TX |
| Opto 22 (Manufacturer) | Temecula, CA |
| Quantum Research Group Ltd. (Manufacturer) | United Kingdom |
| Renesas Technology Corp. (Manufacturer) | Japan |
| Rockwell Automation / Allen-Bradley (Manufacturer) | Milwaukee, WI |
| Samsung Electro-Mechanics (Manuf. & Service) | Korea |
| SANYO Semiconductor Corp. (Manufacturer) | Allendale, NJ |
| Scientific Instruments, Inc. (Manuf. & Distrib.) | West Palm Beach, FL |
| Siargo, Inc. (Manufacturer) | San Jose, CA |
| Sypris Test & Measurement - Magnetics Division (Manuf. & Service) | Orlando, FL |
| Universal Semiconductor, Inc. (Manuf. & Service) | San Jose, CA |
| Zetex Inc. (Manufacturer) | Commack, NY |
| ZMD AG (Manuf. & Service) | Germany |
| ZMD America Inc. (Manuf. & Service) | Cupertino, CA |

## References


1 Bachman, M. MEMS Sensors and Actuators, Introduction. Retrieved Jun. 20, 2005, from EECS179 Class website: < www.li-bachman.net/eecs179/lectures/lecture05.pdf>.

2 Consumer Electronics Glossary - H. Retrieved Jun. 28, 2005, from Radio Shack website: <http://support.radioshack.com/support_tutorials/glossary/glossary-h.htm>.

3 Cort, A. (2002, Jun 05). How Sensors Make Sense. *Assembly Magazine.* Retrieved Jun 09, 2005, from <http://www.assemblymag.com/CDA/ArticleInformation/features/BNP__Features__Item/0,6493,98719,00>.

4 Data Acquisition Glossary. Retrieved Jun. 09, 2005, from United Electronic Industries website: <http://www.ueidaq.com/support/glossary/E/>.

5 Dorf, Richard C., & Eren, Halit (2005). Dorf, Richard: Sensors. *The Engineering Handbook*, 2nd Edition, p.158. Retrieved Jun 16, 2005, from ENGnetBASE database.

6 Encoders Today. Retrieved Jun.14, 2005, from Rockwell Automation website: <http://www.ab.com/encoders/encoderstoday/pdfs/nov01>.

7 Environmental and Safety Compliance Glossary. Retrieved Jun. 20, 2005, from Western New England College website: <http://www1.wnec.edu/esc/index.cfm?selection=doc.1681>.

8 Figliola, R. S. (2000). Operational Modes of Instrumentation. *Measurement, Instrumentation and Sensors Handbook.* Retrieved Jun 16, 2005, from ENGnetBASE database.

9 Flexpoint, Inc. Additional Consumer Applications. Retrieved Jun. 20, 2005, from Flexible Sensor Systems website: <http://www.flexpoint.com/consumer_add_v2>.

10 Goldin, D. S. (1997). Applications of NASA Technology for Biodiversity Conservation. Retrieved Jun. 20, 2005, from website: <http://science.hq.nasa.gov/earth-sun/science/biodiversity/section_b.html>.

11 Glossary. Retrieved Jun.09, 2005, from Focal Point Microsystems website: <http://fpmicro.com/resources/glossary.htm>.

12 Glossary. Retrieved Jun. 28, 2005, from SPORTS INNOVATION OF THE THIRD MILLENIUM website: <http://www.cric-system.com/W.CRIC/doc-en/page09.htm>.

13 Gurley Precision Instruments. Retrieved Jun. 28, 2005, from website: <http://www.gpi-encoders.com/>.

14 Hansman, J. (2000). Characteristics of Instrumentation. *Measurement, Instrumentation and Sensors Handbook.* Retrieved Jun 16, 2005, from ENGnetBASE database.

15 Henkel, Stephanie (2003, Sept). Editorial. *TechWOW!* Vol. 1. Retrieved Jun 20, 2005, from <http://www.sensorsmag.com/techwow/editorial/>.

16 Hogan, B. J. Encoders Conquer New Worlds. *Design News.* Retrieved Jun 09, 2005, from <http://www.designnews.com/index.asp?layout=articlePrint&articleID=CA151226>.

17 Honeywell. Sensing and Control Glossary. Retrieved Jun. 20, 2005, from Honeywell website: <http://content.honeywell.com/sensing/glossary/glossq-z.stm>.

18 Kester, W. Sensor signal conditioning seminar. Retrieved Jun. 20, 2005, from www.analog.com/UploadedFiles/Associated_Docs/5616748499531994843252495435725317649056293583055Fsect1.PDF.

19 Kissell, T. (2000). *Industrial Electronics*. 2nd ed: Prentice Hall PTR. Retrieved from National Instruments website: <http://zone.ni.com/devzone/conceptd.nsf/webmain/9BCCE7934DACF1298625680700573BC0>.

20 Korte, R. The Basics of Encoders. *Modern Machine Shop Online*. Retrieved Jun 09, 2005, from <http://www.mmsonline.com/articles/049703.html>.

21 Magnetic Encoder Fundamentals. Retrieved Jun. 28, 2005, from National Instruments website: <http://zone.ni.com/devzone/conceptd.nsf/webmain/96E20D06DBAEFF6686256B3500787C20>.

22 Monnin, A. (1998). A New Chapter in Encoder Based Motion Control. Retrieved Jun. 28, 2005, from Sensors Online website: <http://www.sensorsmag.com/articles/1198/enc1198/index.htm>.

23 Nanigian, D. (2004). Some myths and facts about temperature sensing: part ii. Retrieved Jun. 20, 2005, from Industrial Heating website: <http://www.industrialheating.com/CDA/ArticleInformation/coverstory/BNPCoverStoryItem/0,2830,133341,00.html>.

24 Optical Encoder Fundamentals. Retrieved Jun. 28, 2005, from National Instruments website: <http://zone.ni.com/devzone/conceptd.nsf/webmain/73304E4E321EF5EA86256B3500782979>.

25 Rajan, M. (2004). U.S. Industrial Sensor Market to Continue to Grow. Retrieved Jun. 20, 2005, from Business Communications Company, Inc. Research website: <http://www.bccresearch.com/editors/RGB-200N.html>.

26 Roffe, B. (2004). Side Impact Sensor Systems. Retrieved Jun. 20, 2005, from FreshPatents.com website: <http://www.freshpatents.com/Side-impact-sensor-systems-dt20040923ptan20040183287.php>.

27 Sensor. Retrieved Jun. 20, 2005, from Wikipedia website: <http://en.wikipedia.org/wiki/Sensor>.

28 Sensors and accessories. Retrieved Jun. 20, 2005, from x10 Security Accessories website: <http://www.x10.com/security/security_accessories.html>.

29 Sensor fundamentals. Retrieved Jun. 09, 2005, from University of Missouri-Rolla website: <http://web.umr.edu/~liou/ME355/Sensors.html>.

30 Smith, R.L. (2000). Sensors. *The Electrical Engineering Handbook*. Retrieved Jun 16, 2005, from ENGnetBASE database.

31 Snow, P. (1999, Oct 18). Encoding standards. *Electronics News*. Retrieved Jun 09, 2005, from http://www.reed-electronics.com/electronicnews/index.asp?.

32 What is a Sensor? Retrieved Jun.09, 2005, from Amir Bahadori website: <http://www.amirbahadori.com/nerdporn/techresearch/>.

33 What are Sensors? Retrieved Jun. 09, 2005, from National Polar-orbiting Operational Environmental Satellite System website: <http://www.ipo.noaa.gov/Technology/whatAreSensorsPopUp.html>.

34 Wilcoxon Research, Inc. Frequently asked questions. Retrieved Jun. 20, 2005, from http://www.wilcoxon.com/knowdesk_faqs.cfm.

# Electromechanical

# 24 Cooling Equipment

Fans and electronic cooling equipment exist for the sole purpose of reducing the temperature of and around electronic components to improve lifespan, prevent burnouts, and ensure efficiency. To appreciate the necessity for cooling, a basic understanding of thermodynamics is very useful. The second law of thermodynamics states that heat energy always moves from an area with more heat to an area with less heat. Heat moves from hot to cold. When an electronic component gets hot, from current or resistance for example, heat from the component is released into the surrounding air. Cooling systems use this natural law to be effective. While the electronics industry features many forms of cooling and temperature regulation, this chapter will discuss:

- Cooling Fans
- Enclosure Air Conditioners
- Heat Exchangers
- Heat Sinks
- Thermoelectric Coolers

Cooling fans are one of the simplest and cost-effective methods of electronic cooling. **Cooling fans** consist of a motor—usually DC—a mounting bracket, and fan blade fixture. The motor turns the blade fixture, moving air parallel to the axis of rotation. When selecting a cooling fan, buyers may want to take into account airflow, voltage and current ratings, noise and speed. Airflow is how much air the fan moves over time and is usually rated in cubic feet per minute. Voltage and current ratings are rated in Volts and Amperes respectively, noise is rated using adjusted decibels (dBA), and speed is rated in revolutions per minute (RPMs).

**Enclosure air conditioners** are simply refrigeration units similar to conventional air conditioners in that they use similar processes to remove heat from air and move the cooled air to a specific area. In this case, the specific area would be a cabinet or room in which electronic hardware is stored. Air conditioners are usually set to a desired temperature using a thermostat or automation systems.

**Heat exchangers** transfer heat from one fluid or gas to another fluid or gas without mixing the two together. A heat exchange system can be liquid-to-liquid, liquid-to-air, or air-to-air. A simple example of a liquid-to-air heat exchanger is an automotive radiator. Hot water is pumped through the radiator tubing, which has a fan blowing on it[3]. The heat from the water is then absorbed into the air being blown across the tubing. Similar systems can be found in electronics when extensive cooling is required, as this is clearly an effective method. Heat exchangers require additional maintenance due to pumps, tubing, motors, and other components involved, so the benefits and costs should be measured before applying such a system[3].

A **heat sink** is a piece of metal that is thermally conductive, designed in a way that allows it to have a large surface area through the use of "fins." The larger surface area created by these fins enables air to come into contact with more of the heat sink material[4]. Therefore, when the heat sink gains heat from a component, the heat is exposed to the air surrounding the fins and the heat is then dissipated, or lost, to the air. This results in the air gaining the heat from the component that is generating the heat. An example of a heat sink is any central processing unit (CPU) made today. The heat generated by modern circuits requires not only heat sinks but cooling fans to move fresh, cool air to the hot heat sink. Heat sinks have no moving parts and are simply "pieces of metal" that play an integral role in electronic cooling[4].

Thermoelectric coolers are one of the newer and more interesting methods of electronic cooling. A **thermoelectric cooler** is a solid-state device that uses a thermoelectric effect and semiconductor materials to function[5]. These coolers are often made of two ceramic plates separated by semiconductor materials. The behavior of the cooler is fairly complex and involves understanding P- and N-type semiconductors. When a thermoelectric cooler is hooked up to a DC voltage source, one side of the plate loses heat, or cools down, while the other side gains heat, or heats up. In applications, a thermoelectric cooler's "cold" side is attached to whatever needs to be cooled and then hooked up to a power source. The heat from the component being cooled is transferred through the cooler and radiated off of the "hot" side[5].

# The Product

## Applications and End Uses

Cooling fans are typically used to cool computers and other home appliances. They are also used in telecommunications, military, and general industrial applications. Additionally, fans can be observed in heating ventilation and air conditioning, or (HVAC), systems[1].

Enclosure air conditioners are typically found in cabinets or rooms in which electronic hardware is stored. The common hardware found in cabinets using enclosure air conditioners is used for, but not limited to, networking, communication, and industrial applications[2].

Heat exchangers are most frequently found in automotive radiators, which are liquid-to-air heat exchangers. Other places heat exchangers are located include boilers, furnaces, refrigerators, and air conditioning systems. Actually, each air conditioning system and refrigeration system has at least two heat exchangers, one on the cooling side and one to eject the heat[3].

Heat sinks are used to cool microprocessors and other chips. Usually, they are glued or clipped to the top or side of a chips package[4].

## Product Features and User Benefits

**Figure 1:** Product Features and Respective End-User Benefits for Select Cooling Products

| Product Features | | End-User Benefits |
|---|---|---|
| Quiet Functioning | ⇨ | Low noise output during functioning allows implementation in situations sensitive to high noise output (e.g., offices, work rooms). |
| Low Power Requirements | ⇨ | Low voltage requirement allows multiple products to be implemented together, making effective use of allotted power supply. |
| Long Life | ⇨ | Product will not wear out prematurely. Results in reduction in frequency of maintenance and/or replacement, which provides reduced system downtime. |
| Variable Output | ⇨ | Heat reduction capacity of device can be varied based on required level of performance. Results in device being active when required and inactive or in a lower power-usage state. |
| High Output | ⇨ | Device is capable of cooling components better, effectively reducing the number of cooling devices required or lowering the cooling requirements of existing devices. |

## Life Cycle

At this stage of technological development, heat sinks and cooling fans are both solidly placed in the maturity phase of the product life cycle. Almost every processor a PC uses has a heat sink attached to it with a cooling fan attached to the heat sink, which some refer to as a **fan sink**. The effectiveness of this combination, coupled with the ease of implementation, makes it very appealing.

Enclosure air conditioners, while effective at lowering cabinet temperature, do not have the "direct" cooling effects provided by the other cooling methods mentioned. Other reasons include efficiency, noise, and scalability. Air conditioners add an additional level of cooling to systems already being cooled through other means, greatly reducing the chance of overheating.

Heat exchangers are used in high-performance situations in which heat sink and fan cooling is not adequate. Used in cases that require large amounts of cooling, heat exchangers are often found in the liquid-to-air form. Cooling requirements have not gone so far as to require that heat exchangers be commonplace, although they are still used. Their use will likely increase in the coming years if computing technology continues to surpass cooling technology.

Thermoelectric coolers are extremely precise devices and, due to current limitations and design, can be a bit expensive and thus only used in situations requiring strict cooling specifications. The relative price and market penetration of thermoelectric coolers puts it, for now at least, in the growth stage of the product life cycle. In coming years, should technology advance to the point where they can be less expensive, more efficient, and/or more versatile, they will likely take the place of mainstream cooling technology, such as cooling fans.

**Figure 2:** Life Cycle Curve for Select Electronic Cooling Methods.

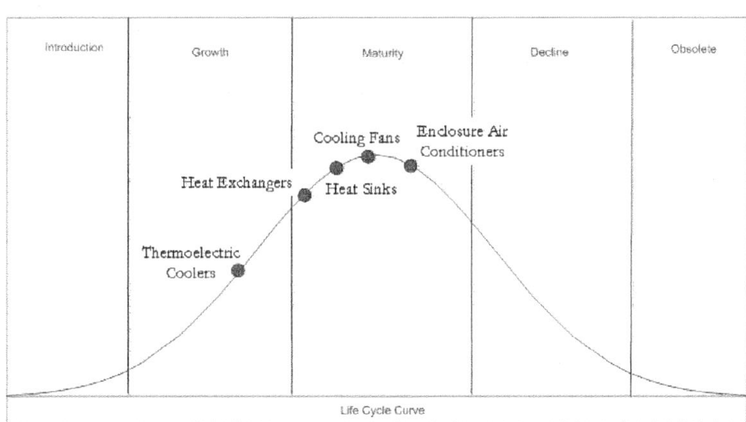

# The Market

## Market Trends

While forecasts for the future market of thermal management products are limited to annual growth percentages and extrapolation, it is clear that an increase in hardware capabilities will be accompanied by an increased need for cooling systems and devices more efficient and effective than current products[6].

The electronic cooling market has been rekindled with new designs that require new cooling solutions. New components and new materials will spark further growth as cooling systems and devices become more efficient and more of a requirement[6].

**Figure 3:** TAM for Electronic Cooling/Thermal Management[6]

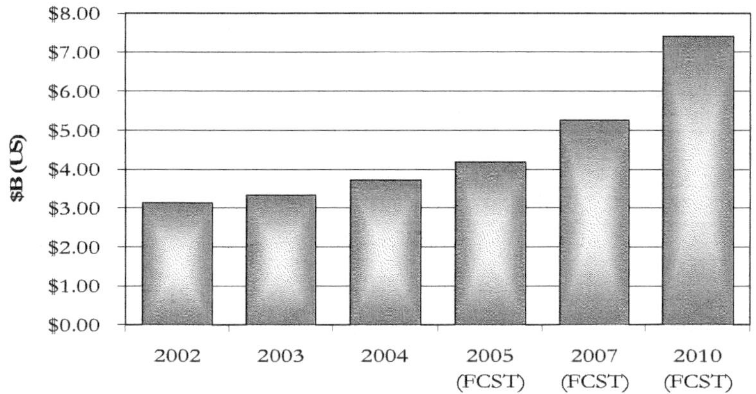


| 2002 | 2003 | 2004 | 2005 (FCST) | 2007 (FCST) | 2010 (FCST) |
|---|---|---|---|---|---|
| $3.13B | $3.32B | $3.73B | $4.18B | $5.25B | $7.40B |

# The Players

## Key Players

In the thermal management sector, many companies exist that are capable of providing the cooling solutions required for a particular project or system.

**Figure 4:** Various Manufacturers, Their Websites and Distributors

| Supplier | Website | Distributors |
|---|---|---|
| Comair Rotron | www.comairrotron.com | Arrow; Avnet; Allied Digi-Key; Newark InOne; RS Electronics |
| EBM-Papst | www.ebmpapst.com | Arrow; Allied; Digi-Key; Mouser |
| NMB Technologies | www.nmbtc.com | Arrow; El-Com; Hammond; Leeds; Newark InOne; RPC Electronics; Sager |
| Panasonic | www.panasonic.com | Arrow; Avnet; Digi-Key; ElectriTek; Norvell; Sager |
| Micronel | www.micronel.ch | |
| Orion | www.orionfans.com | Allied; Carlton-Bates; Mouser; Powell |
| ADDA | www.adda.com.tw | Mouser |
| Sunon | www.sunonusa.com | Allied; Digi-Key; PSC |
| Aavid Thermalloy | www.aavidthermalloy.com | All American; Allied; Arrow Avnet; Digi-Key; Force; Future; Mouser; Newark InOne; Norvell; Sager |

## References


1 About Electronic Cooling Fans. (2005). Retrieved May 5, 2005 from Global Spec website: <http://electronic-components.globalspec.com/LearnMore/Electrical_Electronic_Components/Fans_Electronic_Cooling/Electronic_Cooling_Fans_Blowers>.

2 About Enclosure Air Conditioners. (2005). Retrieved May 5, 2005 from Global Spec website: <http://electronic-components.globalspec.com/LearnMore/Electrical_Electronic_Components/Fans_Electronic_Cooling/Enclosure_Air_Conditioners>.

3 About Heat Exchangers. (2005). Retrieved May 5, 2005 from Global Spec website: <http://http://heat-exchangers.globalspec.com/LearnMore/ Electrical_Electronic_Components/Fans_Electronic_Cooling/Heat_Exchangers>.

4 About Heat Sinks. (2005). Retrieved May 5, 2005 from Global Spec website: <http://electronic-components.globalspec.com/LearnMore/ Electrical_Electronic_Components/Fans_Electronic_Cooling/Heat_Sinks>.

5 Noll, S.J. (2000). Peltier Device Information Directory. Retrieved May 5, 2005 from Peltier website: <http://www.peltier-info.com/>.

6 Ravi, Krishnan. (2003). The Market for Electronics Thermal Management Technologies. GB-185N. Retrieved from BCC Research.

# 25 Relays

A **relay** is an electromechanical device that uses a small voltage to actuate a "switch." Basic relays consist of a coil or electromagnet, a spring armature, an energizing circuit, and a contact circuit. The armature rests near the coil and is wired into the contact circuit. The coil is wired into the energizing circuit. When sufficient current flows through the energizing circuit, a magnetic field is generated and the armature becomes drawn to the coil. When the armature moves, it closes the contact circuit and allows the contact current to flow through the closed switch. When the energizing current stops flowing, the spring pulls the armature back, opening the contact circuit and preventing current from flowing. Relays can be designed so that they are **normally open** (NO), meaning that when an energizing current exists, the armature will close, or **normally closed** (NC), in which case the armature is closed until an energizing current flows through the coil[1].

Another relay type is called a **solid-state relay** (SSR). SSRs are complex semiconductor devices that, as the name suggests, are "solid" and have no moving parts. The relay function is performed using semiconductors and sometimes optical components[2].

## The Product

### Applications and End Uses

Both electromechanical and solid-state relays are used in many applications, specifically in situations that require electronic control of systems with large or even dangerous loads. Relays are also used to implement basic logic functions.

Electromechanical and solid-state relays are used in many of today's automotive, industrial, military/aerospace, and communication applications. Relay mounts come in numerous varieties depending on where they are used. Some relays are still mounted to bases that have the contacts corresponding to the various contacts on the relay, while others are now mounted directly to printed circuit boards (PCBs). In 2004, electromechanical relays made up approximately 79% of the global relay market revenue, while solid-state relays made up only 13%[3].

## Product Features and User Benefits

**Figure 1:** Product Features and User Benefits of Relays

| Product Features | | User Benefits |
|---|---|---|
| Low Contact Resistance | ⇨ | Creates low power consumption that results in longer battery life and allows power to be used elsewhere. |
| Fast Switching Time | ⇨ | Relay is capable of rapid switching, both to and from its actuated position, resulting in precise control in time-sensitive situations. |
| Low Temperature | ⇨ | When used over extended periods of time, does not generate large amounts of heat; results in smooth operation and reduction of additional cooling equipment. |
| Compact Size/Small Footprint | ⇨ | Relay has small dimensions, allowing for board "real estate" to be either reduced in design, making the board smaller, or to be used more efficiently with other components. |

## Life Cycle

Due to heavy use in the automotive and industrial sectors, electromechanical relays are still situated at peak maturity, meaning that their use is still widespread and continuing. While they may be replaced in the future, electromechanical relays will most likely not slip from the maturity stage for many years to come. It is important to note, however, that solid-state relays have begun replacing electro-mechanical (EM) relays. This replacement will continue as solid-state relays improve in performance and as their prices decline to price parity with their electro-mechanical alternative.

**Figure 2:** Life Cycle Curve for Electromechanical and Solid-State Relays

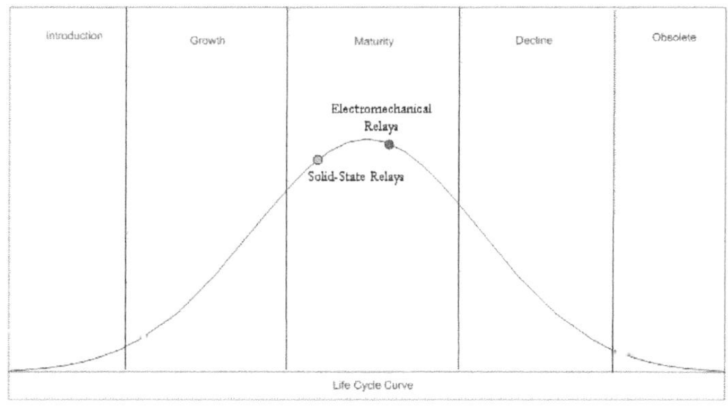

# The Market

## Market Trends

From 2000 to 2002, the global market for relays was in the mid- to high-$4 billion range, while 2002 saw a sharp decline in relay sales along with the trend experienced throughout the electronics industry. Fortunately, if the market maintains the forecasted annual growth rate of approximately 4.2%, 2010's forecast will surpass 2000's $4.75 billion with an estimated $4.98 billion in revenue worldwide[3;6].

**Figure 3:** TAM for Global Relay Market[3;4; 6]

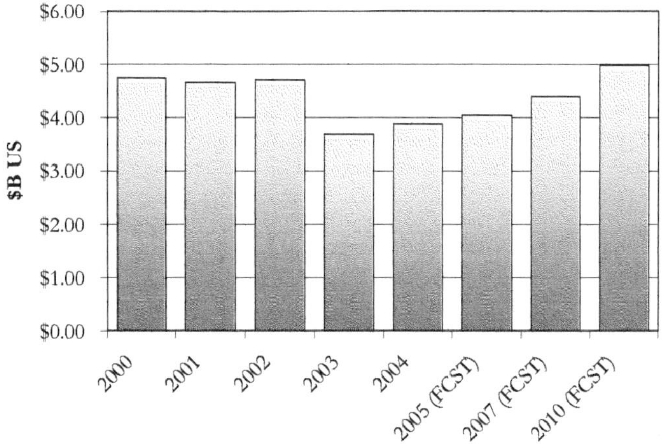


| 2000 | 2001 | 2002 | 2003 | 2004 | 2005 (FCST) | 2007 (FCST) | 2010 (FCST) |
|---|---|---|---|---|---|---|---|
| $4.75 | $4.66 | $4.71 | $3.68 | $3.89 | $4.05 | $4.40 | $4.98 |

## Distributor TAM

In 2003, an estimated 86.1% of U.S. switch, relay, and other electromechanical components were sold through distribution. This high percentage is likely due to the commodity status of most relays as well as other electromechanical components[5].

# The Players

## Key Players

**Figure 4**: Select Relay Suppliers, Their Websites and Distributors

| Supplier | Website | Distributors |
|---|---|---|
| Omron | www.omron.com | Arrow; Avnet; Carlton-Bates; Grainger; Newark InOne; Sager; TTI; Wes Garde |
| Aromat (Matsushita) | www.aromat.com | *As of June 2005, is no longer a supplier to the United States. |
| Tyco Electronics | www.tycoelectronics.com | Arrow; Avnet; Carlton-Bates; Digi-Key; Future; Mouser; TTI; Sager |
| Fujitsu | www.fcai.fujitsu.com | Arrow; Avnet; Force; Jaco; Mouser |
| CII | www.ciitech.com | * Products carried by Tyco Electronics |
| Teledyne | www.teledyne.com | Allied; ASAP; Avnet; Sherburn |
| Sumida | www.sumida.com | Avnet; Digi-Key; Future; Harvey King; Inductors |

## References

1 About Electromechanical Relays. (2005). Retrieved May 10, 2005 from Globalspec website: <http://relays.globalspec.com/LearnMore/Electrical_Electronic_Components/Relays_Timers/Electromechanical_Relays>.

2 About Solid-State Relays. (2005). Retrieved May 10, 2005 from GlobalSpec website: <http://relays.globalspec.com/LearnMore/Electrical_Electronic_Components/Relays_Timers/Solid_State_Relays>.

3 Barnard, R. (2005). The 2004 Global Electromechanical and Solid-State Relay Market Intelligence Program. *Venture Development Corporation.*

4 Chin, S. (2002). Bracing For a Comeback – Relay and switch makers are arming themselves with service programs and fresh products in an effort to counter falling prices and a new wave of competition. Retrieved May 12, 2005 from ProQuest.

5 Gawel, R. (2003). Distribution survey results: Informative distributors seize the day. *Electronic Design.* Retrieved May 12, 2005 from ProQuest.

6 Gordon, J. (2002). 2001/2002 Global Relay Market Intelligence Service. Retrieved May 2005 from Venture Development Corporation.

# 26
# Switches

A **switch** is a simple device that has one universally definitive trait: It allows electric current to flow when closed and prevents current from flowing when opened. While a device may only have to satisfy one general requirement to be a switch, multiple types of switches are available that fit specific requirements of an electronic system. Most switches are split into at least three categories:

- Subminiature
- Miniature
- Standard or Industrial

While there are no standards for sizes in the switch industry, they tend to be classified by size and function of the switch. A number of categories of switches are on the market. However, the eight most commonly used include: Dual In-Line Package, Keylock, pushbutton, rocker, slide, snap-action, tact, and toggle.

## The Product

### Applications and End Uses

**Dual In-line Package, or DIP**, switches are simply a single package with a row of tiny switches. DIP switches can be arrays of tiny rocker, slide, or side-actuated lever ("piano") switches[1]. The switches can have three positions and are used to configure computer peripherals, such as modems and printers. DIP switches made up 5.8% of 2001's global switch revenue[11].

**Keylock** switches are actuated by inserting a key into the switch and turning. Like most switches, keylock switches come in subminiature, miniature, and standard/industrial sizes. Keylock switches have two main attributes—angle between positions and mechanical life[2]. The angle is intuitive in that a four-position switch would have 90 degrees between positions, a 12-position switch would have 30 degrees, etc. An important aspect to understanding keylock switches is that the electrical life expectancy is usually less then the mechanical life expectancy. This should be taken into account when choosing types of keylock switches. Keylock switches make up 1.7% of global switch revenue[11].

Switches classified as **pushbutton (PB)**, have a mechanical plunger that is pressed to actuate the switch and open or close the electrical circuit, depending on if the switch is normally open (NO) or normally closed (NC). PB switches come in subminiature, miniature, and standard sizes. The larger sizes can also be illuminated or non-illuminated[3;16]. In 2001, PB switches made up 21.8% of global switch revenue—the largest category[11].

A **rocker** switch has an actuator that, when actuated, resembles the movement of a rocking chair, hence the name. Rocker switches have two main categories: momentary contact and maintained contact. Momentary contact switches are brought to their actuated state only when a force is applied to the switch. When the force is removed, the switch returns to its normal position. Maintained contact switches are brought to their actuated state simply by a press of the actuator. The switch then enters its actuated state without further force[4]. Many rocker switches used in commercial electronics are marked with an "O" on the side representing "open" or "off" and a "–" on the side representing "closed" or "on." While many rocker switches are only "on and off" switches with two positions, there are also three-position rocker switches available in both momentary and maintained contact categories. In 2001, rocker switches contributed to 10.7% of global switch revenue[11].

**Slide** switches are named after the sliding actuator used to bring the switch to its actuated position or positions. Slide switches, like rocker switches, have momentary- and maintained-contact categories as well as three-position varieties. Momentary-contact switches require a constant force to stay actuated, while maintained-contact switches simply require the initial actuation[5]. Slide switches are found in many end-use applications, such as selector switches on portable audio devices and voltage selector switches on test equipment. In 2001, slide switches made up 8.1% of global switch revenue[11].

**Snap action** switches are mechanical switches that produce a very rapid transfer of contacts from one position to another, making them useful in situations requiring fast opening or closing of a circuit, such as a mouse button. The three kinds of actuators for snap action switches are: lever, pushbutton, and roller[6]. The three classifications of snap action switches are: subminiature, miniature, and standard[9]. Snap action switches are widely used and they made up 18.8% of the 2001 global switch revenue[11].

**Tact** switches are designed to have a small actuating force and are commonly found in keyboards, keypads, and control panel interfaces[19]. Tact switches come in **through-hole and surface-mount** types. Through-hole types have pins that are designed to go through holes on a printed circuit board (PCB) and then go through the soldering process to tie it in to the PCB circuit. Surface-mount types have leads that are mated to a pad already mounted on a PCB[8]. Tact switches are by far the most popular switch in terms of unit sales but don't contribute significantly to total revenue due to the low average unit cost— approximately one-seventh the cost of snap-action switches, the next lowest-priced switch[16]. In 2001, tact switches made up 11.3% of global switch revenue[11].

**Toggle** switches, like slide switches, also come in both maintained- and momentary-contact types. In addition, toggle switches come in subminiature, miniature, and standard sizes. A toggle handle is actuated to open or close the circuit and toggle handles can have various lengths, colors, and options, such as LED-lit[7].

## Product Features and User Benefits

The features and benefits of switches are what differentiate a commoditized product to buyers. Listed below are some general features of switches and their respective user benefits.

**Figure 1:** Features and Respective User Benefits for Electronic Switches

| Product Features | | User Benefits |
|---|---|---|
| Sealed | ⇨ | A barrier is built into the switch to separate the internal contacts and components from the outside world. This helps to keep dust, moisture, and other contaminants out of the switch, extending the switch's functional life by protecting internal moving parts and contacts. |
| Reduced Footprint | ⇨ | Switch takes up less area on PCB, enabling the final product's size to be reduced or enabling more components to fit into the same area. |
| Illuminated Button Face | ⇨ | Button face is backlit with a light-emitting diode (LED) or filament bulb, allowing users to see the button in total darkness[13]. |
| High Life Expectancy | ⇨ | Switch functions for thousands to tens of thousands of operations and more, resulting in a lower frequency of replacement and a more reliable switch with a longer life. |
| Size Variability | ⇨ | A given switch with a desired function can usually be purchased in a range of sizes, allowing designers to match the appropriate switch to the product. Common sizes are subminiature, miniature, and standard. |
| Color-Coded | ⇨ | Switch is given a distinct color, through a light or colored composition, and can therefore be rapidly identified in time-sensitive situations. |
| Light Operating Pressure | ⇨ | Switch can be actuated easily, resulting in simpler operation of device as well as reduced chance of switch damage. Also makes switch more suitable for instrumentation applications[14]. |
| Mounting Options | ⇨ | Switch is mountable onto PCBs via through-hole leads or surface-mount leads. |

## Life Cycle

While most switches are clustered around the peak maturity stage of the product life cycle, two switches stand out as being on the decline. DIP and slide switches are slowly being replaced. DIP switches, originally used to specify addresses and settings to hardware, are being phased out and replaced by solutions that don't require a specific hardware and can be adjusted through software means. Slide switches have also "fallen out of favor"[16] in the switch industry and are simply not used as much as they were in years past.

Another switch of note on the curve is the tact switch. While not "new" per se, the tact switch is the newest switch available of the eight discussed and is finding a number of applications in the industry as discussed earlier in this chapter.

**Figure 2:** Life Cycle Curve for Select Electronic Switch Types

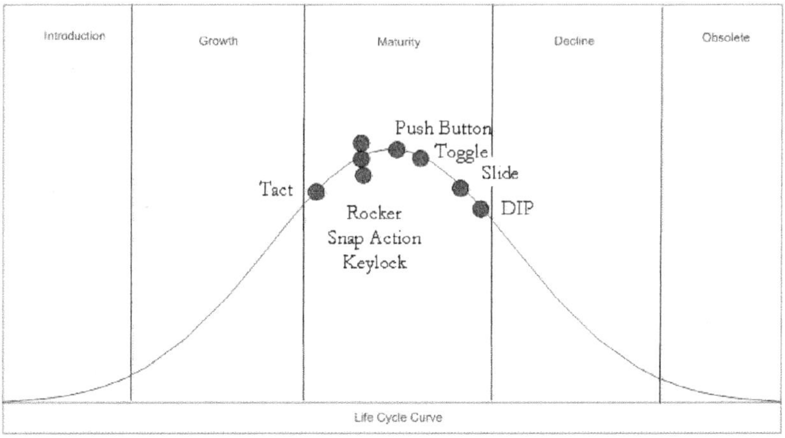

## Market Trends

A look at the 2003 data for the eight switches discussed earlier in the chapter doesn't identify any single switch as being dominant. However, when unit sales are taken into account, tact switches lead every other switch, securing approximately 65.5% of reported switch sales in North America[8].

**Figure 3:** 2003 Estimated Percent Market Share of Revenue for Select Electronic Switches by Type[8].

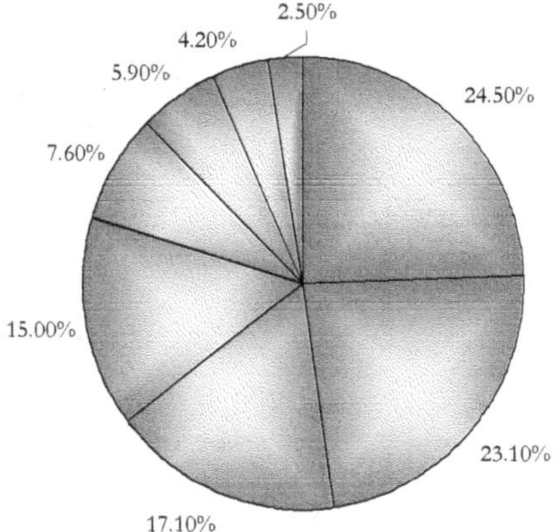


| 2003 Market Share of Revenue for Select Switches | |
|---|---|
| Rocker | 24.5% |
| Snap Action | 23.1% |
| Tact | 17.1% |
| Push Button | 15.0% |
| DIP | 7.6% |
| Toggle | 5.9% |
| Slide | 4.2% |
| Keylock | 2.5% |

In past years, the market has not seen many dramatic changes in global market dollars but forecasts suggest a slow and steady growth, with 2010 reaching the $4 billion mark[10]. Average sale prices (ASPs) for switches in 2004 are lower than the same quarter's ASP in 2003, indicating that revenue is possibly lower and costs may have to be recouped elsewhere in the pricing model of global suppliers[9].

**Figure 4:** Global TAM for Electronic Switches[10;11;12]

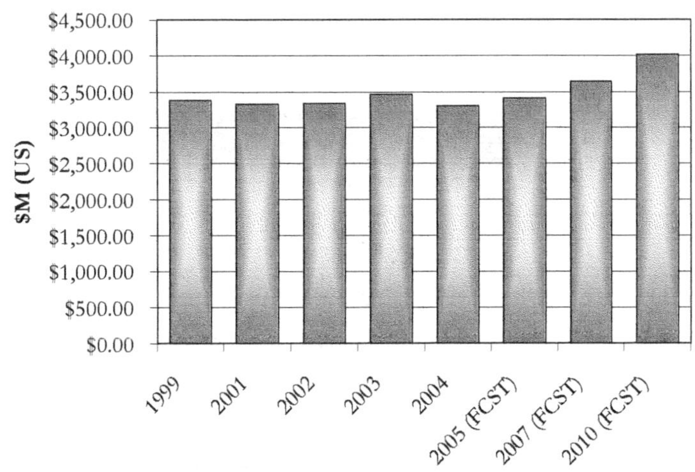


| 1999 | 2001 | 2002 | 2003 | 2004 |
|---|---|---|---|---|
| $3,385 | $3,325 | $3,336 | $3,464 | $3,300 |

| 2005 (FCST) | 2007 (FCST) | 2010 (FCST) |
|---|---|---|
| $3,409 | $3,638 | $4,010 |

## Distributor TAM

Capturing approximately 40% of 2001's U.S. switch sales, distributors are continuing to function as the primary outlet for U.S. switch sales. The percentage today is higher and will likely keep growing for some time due to the market behavior in which commodity products, such as switches, are typically sold through distribution. The percentage moving through distribution will also be increased due to suppliers looking to distributors for their expertise in forecasting, inventory management, and fee-based services[18].

## Market Segmentation

**Figure 5:** 2003 Market Segmentation for Electronic Switches[15]

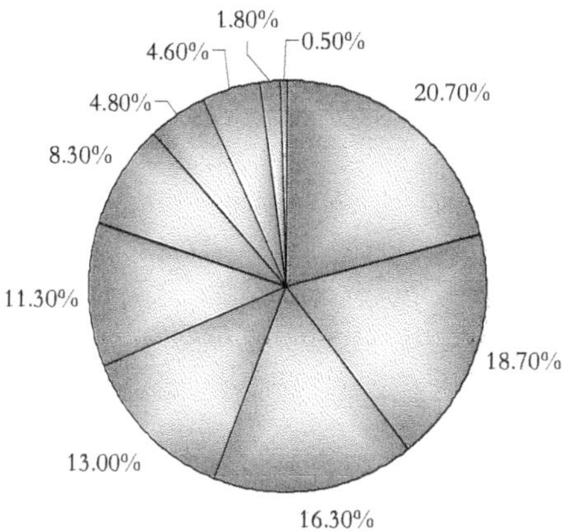


| Market Segmentation for Electronic Switches | |
|---|---|
| Industrial Machinery, Automation, & Process Controls | 20.70% |
| Telecom, Cable, & Broadcast Equipment | 18.70% |
| Computers, Peripherals, & Office Equipment | 16.30% |
| Commercial Equipment | 13.00% |
| Consumer Electronics | 11.30% |
| Appliances | 8.30% |
| Non-Automotive Vehicles | 4.80% |
| Medical Equipment | 4.60% |
| Test & Laboratory Equipment | 1.80% |
| Power Tools & Lawn/Garden Equipment | 0.50% |

# The Players

## Key Players

**Figure 6:** Selected Suppliers of Switches

| Supplier | Website | Distributors |
|---|---|---|
| Alps | www.alps.com | Bisco Industries; Mouser |
| Carling Technologies | www.carlingtech.com | Allied; Carlton-Bates; Digi-Key; Future; Mouser; Newark InOne; Sager |
| Cherry Corp | www.cherrycorp.com | Avnet; Digi-Key; Sager |
| Eaton | www.eaton.com | Summit |
| Honeywell | www.honeywell.com | Allied; Arrow; RS Electronics; Sager |
| Idec Izumi | www.idec.com | Allied; Carlton-Bates; Newark InOne |
| ITT Cannon/ C&K | www.ittcannon.com | Avnet; Allied; Arrow; Carlton-Bates; Digi-Key; Future Electronics; Mouser; Newark InOne; Sager |
| Marquardt | www.switches.com | Force; Nedco; Wes Garde |
| NKK | www.nkkswitches.com | All American; Allied; Arrow; Carlton-Bates; Future; Mouser |
| Omron | www.omron.com | Arrow; Avnet; Carlton-Bates; Grainger; Newark InOne; Sager; TTI; Wes Garde |
| Saia-Burgess | www.saia-burgess.com | Beyond; Esco; Nedco; RSI |
| Siber | www.siberam.com | Bergquist; Carlton-Bates; PSC; Wes Garde |
| Tyco Electronics | www.tycoelectronics.com | Arrow; Avnet; DAC; Sager; TTI |

## References


1 About DIP Switches. (1998). Retrieved Apr. 19, 2005 from GlobalSpec website: <http://switches.globalsepc.com/LearnMore/Electrical_Electronic_Compnents/Switches/DIP_Switches>.

2 About Keylock Switches. (1998). Retrieved Apr. 19, 2005 from GlobalSpec website: <http://switches.globalspec.com/LearnMore/Electrical_Electronic_Components/Switches/DIP_Switches>.

3 About Pushbutton Switches. (1998). Retrieved Apr. 19, 2005 from GlobalSpec website: <http://switches.globalspec.com/LearnMore/Electrical_Electronic_Components/Switches/Pushbutton_Switches>.

4 About Rocker Switches, Maintained Contact. (1998). Retrieved Apr. 19, 2005 from GlobalSpec website: <http://switches.globalspec.com/LearnMore/Electrical_Electronic_Components/Switches/Rocker_Switches>.

5 About Slide Switches, Maintained Contact. (1998). Retrieved Apr. 19, 2005 from GlobalSpec website: <http://switches.globalspec.com/LearnMore/Electrical_Electronic_Components/Switches/Slide_Switches_Maintained_Contact>.

6 About Snap Action Switches. (1998). Retrieved Apr. 19, 2005 from GlobalSpec website: <http://switches.globalspec.com/LearnMore/Electrical_Electronic_Components/Switches/Snap_Action_Switches>.

7 About Toggle Switches, Maintained Contact. (1998). Retrieved Apr. 19, 2005 from GlobalSpec website: <http://switches.globalspec.com/LearnMore/Electrical_Electronic_Components/Switches/Toggle_Switches>.

8 Cumulus. (2004). Switch Tracks: North American 2003 End-User, Sales Channel & Geographic Report. Retrieved from Cumulus, Inc.

9 Cumulus. (2005). Switch Tracks: North American Issue No. 9, Volume 1 1Q05. Retrieved from Cumulus, Inc.

10 Good news: Switches will remain a buyer's market. (2005). *Purchasing Magazine*. Retrieved Apr. 25, 2005 from Purchasing website: <http://www.purchasing.com/article/CA510896.html>.

11 Gordon, J. and Millette, J. (2002). 2002 Global Switch Market Intelligence Service: An Executive White Paper for Senior Managers Interested in Maximizing Growth and Profitability. Retrieved from Venture Development Corporation.

12 Gordon, J. (2002). VDC Finds Growth in Switch Consumption Coming from 3 Regions, in 3 Phases, in 2003. Retrieved from Venture Development Corporation.

13 LED Features and Benefits. (1996). Retrieved Apr. 23, 2004 from Vivisun website: <http://www.vivisun.com/products/led_features.html>.

14 Pushbutton Switches by Eaton Commercial Controls Division Advertisement. (n.d.). Retrieved from Eaton Corporation website: <http://comericalcontrols/eaton.com/VCBU/catalog/commcatl.pdx/Push.pdf>.

15 Schwert, M. (2004). Soft Patch or Leveling Off? Retrieved from TTI, Inc. website: <http://www.ttiinc.com/object/ME_Schwet_20040927>.

16 Schwert, M. (2005). Telephone Interview.

17 Sullivan, L. (2002). Distributor sales of EM switches rising. Retrieved from Electronics Supply & Manufacturing website: <http://www.my-esm.com/story/OEG20020412S0133>.

18 Switches. (1998). Retrieved Apr. 19, 2005 from GlobalSpec website: <http://switches.globalspec.com/ProductFinder/Electrical_Electronic_Components/Switches>.

# Glossary

| | |
|---|---|
| 1 giga-bit | 1 billion bits |
| 1 mega-bit | 1 million bits |
| 1 tera-bit | 1 trillion bits |
| 2G | See Second Generation Products. |
| 3G | See Third Generation Products. |
| 802.11 | A group of specifications for WLAN technology created by the Institute of Electrical and Electronics Engineers (IEEE); it details a connection from a wireless user and a base station or between two wireless users. |
| 802.11a | Part of the 802.11 specification, is a technology that operates at 54 Megabits and it is considered to be "Wi-Fi Certified." |
| 802.11b | Part of the 802.11 specification, is a technology that operates at 11 Megabits and considered to be "Wi-Fi Certified." |
| 802.11g | Part of the 802.11 specification, is a technology that operates at 54 Megabits and considered to be "Wi-Fi Certified." It is compatible with the 802.11a technology and backwards compatible with the 802.11b technology. |
| A/D Converter | See Analog-to-Digital Converter. |
| AC Power Supply | A device that takes an input AC signal and regulates it in order to fit the requirements of the system. |
| AC/DC Power Supply | Accepts an AC current, often the standard wall current, and converts it to a specific DC current. |
| AC-DC Converter | Accepts AC input voltage from a wall outlet and sends DC voltage. |
| Active Filter | A type of analog electronic filter that combines active components, usually operational amplifiers, with resistors and capacitors (passive components). |
| Adaptive Differential Pulse Code Modulation | A process that varies the size of the samples and is able to switch the coding scale to compensate for variations in the advanced technique that does not use absolute measurements for each sample point. It also codes the differences in frequency and amplitude. It is a modification of Differential Pulse Code Modulation. |
| ADC | See Analog-to-Digital Converter. |
| Address Bus | Used for communicating the physical address of computer memory elements and locations that the requesting unit wants to access or read/write. Transfers only information. |
| ADPCM | See Adaptive Differential Pulse Code Modulation. |

| | |
|---|---|
| Alternators | A device that generates a three-phase alternating current. |
| ALU | See Arithmetic/Logic Unit. |
| Aluminum Electrolytic Capacitor | A capacitor that has a dielectric that is formed by an electrolytic process and is classified as either wet or dry. |
| Amplitude Modulation (AM) | Modulation in which the amplitude of a carrier wave is varied in accordance with some characteristic of the modulating sigma. |
| Analog Filter | A filter that uses analog electronic circuits to produce a filtering effect on analog signals, such as electric potential, sound waves, and mechanical motion. |
| Analog-to-Digital Converter | A device that changes continually varying analog signals into a digital signal or binary code that can be used by a computer or other digital devices. |
| Antenna | A device that is conductive and designed to radiate RF energy from a transmitter, or to capture RF energy for application to a receiver. |
| Antifuse | An alternative to fusible-link in which a programming unit is used to apply voltage that results in a connection pattern designed to perform the desired function. |
| Application Specific Integrated Circuit | A chip that is custom designed for a special application performing a fixed set of steps that cannot be changed. (ASIC pronounced 'a-sick') |
| Application Specific Standard Product | A semiconductor device, integrated circuit product that is dedicated to a specific application market and sold to more than one user (standard). |
| Architecture | Refers to the design and technology of the DSP chip. |
| Arithmetic/Logic Unit | Part of a microprocessor or CPU that performs mathematical calculations. |
| ASIC | See Application Specific Integrated Circuit. |
| ASSP | See Application Specific Standard Product. |
| Astable | An electronic circuit that has no stable states and therefore changes back and forth, oscillating between the two unstable states. |
| Astable Multivibrator | Controls the frequency and duty cycle of a timer with two external resistors and a single external capacitor. |
| Asynchronous Counter | A type of counter in which each stage is clocked from the output of the preceding stage. |
| Attenuator | See Fiber Optic Attenuator. |
| Bandwidth (Processors) | The number of bits that are processed in a single instruction. |
| Bandwidth (Filters) | The width of a frequency band measured in hertz. |
| Baud Rates | A measure of the "signaling rate," which is the number of changes to the transmission media per second in a coded (modulated) signal. |

| | |
|---|---|
| Binary Adders | Logic devices that add two binary numbers to form a sum with optional carry. |
| Bipolar Logic Circuits | Logic circuits that developed from discrete components and are components independent of others. |
| Bit-error | Indicates the number of bits in a transmission sequence (flow of data) that has been altered by noise (interference) over a communication channel. |
| Bluetooth | A wireless networking technology that uses radio waves. It is designed to connect and simplify the communications between devices within a short range. |
| Bus | A common pathway, or channel, between multiple devices. |
| Cache | Refers to the collection of data that duplicates original values stored elsewhere or that has been computed earlier. |
| Capacitor | An electrical device designed to store electrical energy. |
| Carrier | Through alternating current vibrations, it establishes a transmission line for a signal to travel. |
| Carrier Wave | An electromagnetic wave that can be modulated— as in frequency, amplitude, or phase—to transmit speech, music, images, or other signals. |
| Cascaded Counters | Meaning connected "end-to-end," as when several counters are connected from the terminal count output of one to the enable input of the next. |
| Cascaded Timers | Meaning connected "end-to-end," as when several timers are connected from the terminal timer output of one to the enable input of the next. |
| CDMA | See Code Division Multiple Access. |
| Cell Library | A specified set of logic functions. |
| Central Processing Unit | See Microprocessor. |
| Ceramic Capacitor | A capacitor whose dielectric is made up of a ceramic material because of its ability to permit electrostatic attraction and repulsion to take place across it. |
| Chassis | Commonly refers to the frame, skeleton, shell, or hull that protects or houses some kind of device. |
| Chip Aluminum Capacitor | A capacitor that does not have any leads and is referred to as a surface mount capacitor. |
| Circular connector | A type of connector used primarily for external device interfacing. |
| Cladding | The coating of glass that surrounds the core of a fiber optic cable. |
| Clamping Diode | A diode used to limit the peak voltage on a line to a pre-determined maximum voltage. |
| Clock Line | Allows a clock pulse sequence in a microprocessor. |

| | |
|---|---|
| Clock Oscillator | An oscillator with no frequency correction system to control variations caused by temperature. |
| Clock Speed | Can determine how many instructions per second the chip can execute, specified in megahertz (MHz). |
| CMOS | A discrete logic family that uses the MOSFET technology. |
| Code Division Multiple Access | A form of WWAN that does not reserve a specific frequency exclusive to the user, but provides access to the full available spectrum. |
| Codec | A device that encodes or decodes a signal through compression and decompression techniques. |
| Comparator (Signal Conditioning) | A device that compares two voltages or currents, and then switches its output to indicate which one of the two voltages or currents is larger. |
| Comparator (Logic) | A logic function that performs magnitude comparisons; it compares two quantities and indicates whether they are equal or not. |
| Connector | The range of electronic products whose sole purpose is to connect devices to one another. |
| Cooling Fans | A device that consists of a motor (usually DC), a mounting bracket, and fan blade fixture. The motor turns the blade fixture, moving air parallel to the axis of rotation. |
| Counter | A digital circuit capable of counting electronic events by progressing through a sequence of binary states. |
| CPLD | Complex Programmable Logic Device |
| CPU | See Microprocessor. |
| CTD counter | Count down counter |
| CTU counter | Count up counter |
| Current Source | Takes an input signal and converts it to a specific AC or DC signal. |
| D/A Converter | See Digital-to-Analog Converter. |
| DAC | See Digital-to-Analog Converter. |
| Data Bus | Used for transferring data requested. |
| Data Converter | A device that changes information from one code to another. |
| Data Interface Receiver | A device that accepts signals and decodes messages. |
| Data Interface Transceiver | A device within one chassis that performs both transmitting and receiving functions that are in a common housing. |
| Data Interface Transmitter | A device that generates and amplifies a carrier wave, modulates it with a meaningful signal derived from speech or other sources, and radiates the resulting signal from an antenna. |
| DC Power Supply | A device that takes an input current and converts it to a desired DC output current. |

| | |
|---|---|
| DC/AC Power Supply | A device that converts an input DC signal into an output AC signal. |
| DC-DC Converter | A device that accepts DC input and provides regulated and/or isolated DC output in various applications. |
| Decoder | An electronic device that separates or combines two or more signals. |
| Demodulator | A device that decodes the modulations; the opposite of what a modulator does. |
| Demultiplexer | A logic circuit that switches digital data from several input lines onto a single output line in a specified time sequence or visa versa. |
| Dielectric | The part of the capacitor that includes two conducting plates separated by an insulator. |
| Differential Pulse Code Modulation | A process that encodes the pulse code modulation values as differences between the current and previous values. |
| Digipot | See Digital Potentiometer. |
| Digital Dividers | Digital circuits that produce the arithmetic operation of binary division. |
| Digital Filter | A filter that performs numerical calculations on sampled values from a signal with the usage of a digital processor. |
| Digital Potentiometer | A potentiometer that has the same functionality as a mechanical potentiometer but uses logical inputs. |
| Digital Signal Processor | A specialized microprocessor designed specifically for converting signals from real-world data into digital data that can be analyzed. |
| Digital-to-Analog Converter | A device that coverts a digital signal into an analog signal. |
| Diode | A simple semiconductor device that allows current to flow in one direction but not the other. It is the electronic version of a one-way valve. |
| Diode Bridge | A special arrangement of four diodes that transforms an alternating current into a direct current, using both positive and negative excursions of a single phase alternating current. |
| DIP Switch | See Dual In-Line Switch. |
| Discrete | Refers to a packaged semiconductor having a single device that has an electrical functional component such as a diode, thyristor, or transistor. |
| Discrete Logic | An individual component or group of components on a chip used to perform one function or set of functions. |
| Discrete Logic Encoder | Electronic devices that separate or combine two or more signals. |
| Discrete Logic Timer | A circuit that can be used as a one shot or as an oscillator. |

| | |
|---|---|
| Distributor Total Available Market | The total world-wide revenues of distributors for the focal product. |
| Distortion | A change in the frequency of the waveform. |
| Double Diodes | A pair of diodes in the same housing, with a split (center-tapped) alternating current supply. |
| DPCM | See Differential Pulse Code Modulation. |
| DRAM | See Dynamic Random Access Memory. |
| Dry Electrolytic Capacitor | A type of aluminum capacitor that has a dry rather than moist electrolyte. |
| DTAM | See Distributor Total Available Market. |
| Dual In-Line Switch | A single package with a row of tiny switches—arrays of tiny rocker, slide, or side-actuated lever ("piano") switches. |
| Dynamic Random Access Memory | A type of random access memory that stores each bit of data in a separate capacitor. |
| Edge Triggered Flip Flop | A type of digital storage circuit in which the entered data appears on the output on the same clock edge. |
| EEPROM | See Electrically Erasable Programmable Read-Only Memory. |
| Electrically Erasable Programmable Read-Only Memory | A form of read-only memory than can be electrically erased and then reprogrammed while in place. |
| Electromechanical | Refers to the use of electricity to run moving parts. |
| Electromechanical Filter | Filters consisting of transducers that convert electrical signal to mechanical motion. |
| Electro-Optical System | A system of devices that electrically isolates two related circuits—usually a transmitter and a receiver—and transfers the power between them. |
| Embedded Controller | A controller that is built or embedded into a greater system that it is designed to help control. |
| Embedded Processor | See Embedded Controller. |
| Enclosure Air Conditioners | Refrigeration units similar to conventional air conditioners that remove heat from air and move the cooled air to a specific area. |
| Encoder | A device that converts data from one form into another. See Discrete Logic Encoder. |
| EPROM | See Erasable Programmable Read-Only Memory. |
| Erasable Programmable Read-Only Memory | Memory that retains its contents until it is exposed to ultraviolet light. |
| Fail-safe | A device that, if it fails, will fail in such a way that it will not cause harm to the device, or system, or at least will minimize the overall effects of failing. |
| Fan sink | A device that is a cooling fan attached to a heat sink. |
| Farad (F) | The unit capacitance is measured in. |

| | |
|---|---|
| FHSS | See Frequency Hop Spread Spectrum. |
| Fiber Optic Attenuator | A device used to reduce the signal power or amplitude in a fiber optic link by causing a loss without distorting the waveform. |
| Fiber Optic Connector | A device that enables the interconnection of separate fiber systems through fiber cables. |
| Fiber Optic Receiver | A device that converts optical signals into electrical signals; it includes a photodiode semiconductor, signal conditioning circuitry, and an amplifier. |
| Fiber Optic Switch | A switch that can transmit an optical signal without the usual electronic-to-optical and optical-to-electronic conversions. |
| Fiber Optic Transmitter | A device that converts electrical signals to optical signals; it includes a source and driving electronics to perform its function. |
| Fiber Optics | The use of glass or plastic fibers to transmit data using light waves. |
| Field Programmable Gate Array | A device consisting of an array of programmable logic blocks that contain logic gates and memory elements. |
| Film Capacitor | A capacitor that has a plastic film dielectric wound alternately with metal foil electrodes. |
| Filter (Filters) | A device that helps separate the unimportant information from the desired information in real-world signals. See Microwave Filter. |
| Fixed Resistor | A resistor with a given value that cannot be changed. |
| Fixed Capacitor | A capacitor with a resistance value that cannot be adjusted. |
| Fixed Point DSP | A DSP that operates on 16-bit words and performs strictly integer arithmetic. |
| Flash Memory | A type of constantly powered, nonvolatile memory that can be erased and reprogrammed in units of memory called blocks. |
| Flip Flop | A basic storage circuit that can store only one bit at a time. Its output can first flip one way and then flop back the other way. |
| Floating Point DSP | A type of DSP that operates on 32- to 40-bit words and performs either integer or real arithmetic. |
| FPGA | See Field Programmable Gate Array. |
| Frequency | The measurement of the number of times that a repeated event occurs per unit of time. |
| Frequency Hop Spread Spectrum | A method of wireless transmission in which a transmitter and receiver hop together from one frequency to another in an arranged pattern. |
| Full Wave Rectifier | See Diode Bridge. |

| | |
|---|---|
| Fuse | A technology that contains tiny fuses that may be programmed and configured to perform almost any logical operation. |
| Fusible Link | See Fuse. |
| GAL | Generic Array Logic |
| Gate | See Logic Gate.<br>See Thyristor Gate. |
| Gate Array | A chip made of basic cells, each containing a number of unconnected transistors and resistors. |
| Global System for Mobile Communication | A software system that is not controlled by any one company, meaning it is open or non-proprietary. |
| Ground Shifts | Vibrations or interference (such as created by machinery) that cause errors in data transferring. |
| GSM | See Global System for Mobile Communication. |
| Half Wave Rectifier | A diode used to convert alternating current electricity into direct current. |
| Half-duplex | The transmission of data in both directions, allowing only one direction at a time. |
| Heat Exchanger | A device that transfers heat from one fluid or gas to another fluid or gas without mixing the two together. |
| Heat Sink | A piece of metal that is thermally conductive, designed in a way that allows it to have a large surface area through the use of "fins." |
| High Refractive Index | Permits light to easily travel throughout the interior or core of a fiber optic cable. |
| High-impedance | High resistance within a circuit against the flow of electrical current. |
| Holographic Memory | A type of optical recording that can have high record density and also maximum data access speed. |
| Home Radio Frequency | Created to provide wireless networks in home, not businesses. No vendors or working groups are supporting HomeRF anymore. |
| HomeRF | See Home Radio Frequency. |
| IC | See Integrated Circuit. |
| Inductor | A passive electrical device made of coiled wire that generates a magnetic field to store energy when current is passed through it. |
| Input Bias | The idle of the device or the power required to keep the device running. |
| Instruction Set | The programmed instructions the chip can execute. |
| Integrated Circuit | Thin chips consisting of at least two interconnected semiconductor devices that are primarily transistors. |

| | |
|---|---|
| Interface | Any point, area, or surface along which two substances meet. Also, a part or IC that sends or receives signals from other parts, ICs, or subsystems. |
| Inverter | The logic circuit that performs the NOT operation. |
| Keylock Switch | A switch that is actuated by inserting a key into the switch and turning it. |
| Kraft Paper | The thin sheet of paper made from wood pulp that is included in paraffin paper capacitors. |
| Laser Diode | A laser with a PN junction in a semiconductor that is used to create analog signals or digital pulses for transmission through optical fibers. |
| Latches | Bistable digital circuits used for storing a bit. They are sets of Flip Flops with a common clock signal. |
| Leaded Capacitor | A capacitor that is connected to a circuit by different kinds of leads. |
| LED | See Light Emitting Diode. |
| Light Emitting Device | An electrical input that produces a light output. |
| Light Emitting Diode | A semiconductor device that emits incoherent monochromatic light when electrically biased in the forward direction. |
| Light Sensing Device | Incoming light that produces an electrical output. |
| Line Driver | A device that extends the transmission distance between terminals and computers. |
| Linear | The relationship between inputs and outputs in which the output varies (changes) in direct proportion to the input. |
| Linear Encoder | An encoder that uses magnetic detectors, optical lenses, inductive technology, or racks and pinions to transfer data. |
| Linear Potentiometers | A potentiometer that, instead of rotating, extends and retracts to control the flow of current. |
| Linear Voltage Regulator | Used to provide constant DC output voltage and contain circuitry that continuously holds the output voltage at the design value regardless of changes in load current or input voltage. |
| Logic Gate | Serves as the building block to digital logic circuits using combinational logical; refers to the small (several micron) structures which take one or more bits as input and produce another bit as output and manipulate 0s and 1s. |
| Logic Product Family | The industry-set subdivision of logic gate production technologies. |
| Low Refractive Index | Does not allow light to pass through and thus results in the reflection back into the core; found in the cladding of a fiber optic cable. |
| Master Slave Flip Flop | A type of digital storage circuit in which the entered data appears on the output on the same clock edge. |

| | |
|---|---|
| Memory | The electronic component used to internally store data. |
| Metal Oxide Silicon Field Effect Transistor | The discrete logic technology that is the replacement to bipolar logic circuits. Its technology reduces space and also draws less power than the bipolar devices. |
| Metal Oxide Varistor | The most widespread type of varistor; a diode junction is formed between each grain in the varistor and these diode junctions allow for current to only flow in one direction. |
| Microcontroller | A single-chip computer used to control a process or function. |
| Microprocessor | An integrated circuit that performs the bulk of the processing in an electronic device. |
| Microwave | An electromagnetic wave that represents a small portion of the wavelength continuum, which can be from 1 millimeter to 30 centimeters in wavelength. |
| Microwave Filter | An electronic device that selectively passes frequency waves from the oscillator into the output. |
| Millions Instructions per Second (MIPS) | The amount of instructions a DSP can process per second. |
| MIPS | See Millions Instructions Per Second. |
| Modulate | To blend a data signal (text, voice, etc.) into a carrier for transmission over a network and to encode information on a carrier wave. |
| Modulator | A circuit or device that combines two different signals in such a way that they can be pulled apart later. Also encodes information on a carrier wave.<br>See RF Modulator. |
| Molecular Memory | A memory system that uses the digital bits of a molecule called bacteriorhodopsin. |
| Monostable | An electronic circuit containing two states, only one of which is stable. |
| Monostable Multivibrator | The timed interval of a timer controlled by a single external resistor and capacitor network. |
| MOSFET | See Metal Oxide Silicon Field Effect Transistors. |
| MOV | See Metal Oxide Varistor. |
| Multi-drop | A connection of several devices to a single communications channel (the path a message follows). |
| Multiplexers | A logic circuit that switches digital data from several input lines onto a single output line in a specified time sequence or visa versa. |
| Multi-turn Potentiometer | A potentiometer that is used when a higher accuracy in adjustment, control, and measuring is desired or when a measurement application requires more than one revolution. |

| | |
|---|---|
| NC | See Normally Closed. |
| Network | A group of computers that communicates through wires or through temporary means like a modem. |
| NIC card | Network Interface Card |
| Niobium Capacitor | A capacitor with a dielectric made of niobium. |
| NO | See Normally Open. |
| Noise | A random change in the frequency. |
| Non-volatile Memory | Memory that retains its contents when power is removed. |
| Normally Closed | The armature is closed until an energizing current flows through the coil. |
| Normally Open | The armature is open until it is specifically closed or energized. |
| Numerical Aperture | The ability to reflect light. |
| OCXO | See Oxygen Controlled Crystal Oscillator. |
| Offset Voltage | The error that is calculated for unbalanced transistors in a circuit. |
| Ohm's Law | Theory that states that the voltage in a circuit is equal to the current times the resistance. |
| One Time Programmable | Once it is programmed it cannot be changed. |
| Op Amp | See Operational Amplifier. |
| Operational Amplifier | Used to boost an A/C signal from its low input level to a higher level that is more easily read by another device. |
| Optically Coupled Pairs | See Optocoupler. |
| Optocoupler | Optoelectronic device that is used to transfer an electrical signal between two circuits while isolating the circuits from each other. |
| Optoelectronic-enabled Product | Any device that uses an optoelectronic component as part of its major function. |
| Optoelectronics | When light (optical energy) is converted into electricity (electrical energy) and when electrical energy is converted into optical energy. |
| Optoisolator | See Optocoupler. |
| Oscillator | An electrical circuit that produces electrical waves at specified frequencies.<br>See RF Oscillator. |
| OTP | See One Time Programmable. |
| Optical Sensors | Sensors that are used to measure velocity and acceleration. |
| Over-current | A current that exceeds the amperage rating of the conductors (cable), equipment (motors, instruments), or other devices in a circuit. |
| Over-voltage | A power surge that takes place when the voltage is 110% above rated RMS voltage for one or more cycles. |

| | |
|---|---|
| Oxygen Controlled Crystal Oscillator | An oscillator that uses temperature control circuitry to keep it at a precise temperature. |
| PAL | Programmable Array Logic |
| Paraffin Paper Capacitors | A capacitor that has a thin sheet of Kraft paper rolled between two metal foils. |
| Parallel Binary Adders | Logic devices that add two binary numbers to form a sum with optional carry. |
| Passive Filter | Considered the simplest form of filters, it is based on their passive components: resistors, inductors, and capacitors. |
| PB | See Pushbutton. |
| PCB Connector | Connectors that are mounted onto printed circuit boards (PCB) using pins on the connector that mate with the holes on the PCB. |
| PCM | See Pulse Code Modulation. |
| PDC | See Personal Digital Cellular. |
| Peripheral | A type of computer hardware that is added to a host computer to expand its abilities. |
| Personal Digital Cellular | A 2G TDMA based protocol that is used in Japan and operates in the 800 and 1500 MHz bands. |
| Photo Coupled Pairs | See Optocoupler. |
| Photocell | A type of resister that allows current to flow freely when light is hitting it; when no light is present, the resistance in the photocell increases, limiting free current flow. |
| Photocoupler | See Optocoupler. |
| Photodiode | An electronic component and a type of photodetector that is a positive-negative (PN) junction designed to be responsive to optical input. |
| Photodiode | A two-electrode, positive-negative (PN) junction that is formed in a semiconductor. |
| Photon Coupled Isolator | See Optocoupler. |
| Photonics | See Optoelectronics. |
| Photophone | One of the building blocks of fiber optics. Developed by Alexander Graham Bell, it carried voices only 200 meters. |
| Phototransistor | An optoelectronic device that is similar to the transistor and also to the photodiode; it is controlled by light hitting the PN junction of the device. |
| Photovoltaic Cell | See Solar Cell. |
| Piconet | A network of devices using Bluetooth technology protocols to allow one master device to interconnect with up to seven slave devices. Also called a Personal Area Network or PAN. |
| Pipelining | When more transistors are used in a microprocessor, the ability for instruction execution to overlap occurs. |
| PLD | See Programmable Logic Device. |

| | |
|---|---|
| Pot | See Potentiometer. |
| Potentiometer | An adjustable resistor that is used to control current flow. |
| Power Conversion Devices | A light input gets converted into electrical power. |
| Power Management IC | See Power Management Integrated Circuit. |
| Power Management Integrated Circuit | A device that regulates the power supplied to another device. |
| Power Supply | An electronic device intended to convert an input electrical current into a different output current. |
| Programmable Logic Device | An integrated circuit that is programmed to perform digital logic and memory functions in a system. |
| Propagation Delay | The time it takes for a change in the input path to create a change in the output path. |
| Pulse Code Modulation | The process that samples an analog signal at uniform intervals; every sample is then quantized into a series of binary code. |
| Pushbutton | A switch with a mechanical plunger that is pressed in order to actuate the switch and open or close the electrical circuit, depending on if the switch is normally open (NO) or normally closed (NC). |
| Radio Frequency | Any frequency within the electromagnetic spectrum associated with radio wave propagation. |
| Rail to Rail | Provides a dynamic range on a low power source, thus allowing the output voltage to be greatly increased. |
| RD Line | See Read Line. |
| Read Line | Copies data to a place in memory where they can be used and stored by a program. |
| Receiver | See RF Receiver.<br>See Data Interface Receiver. |
| Rectangular Connector | A type of connector that includes Universal Serial Bus (USB) and IEEE 1394 Firewire® connectors. |
| Rectifier Diode | A heavy-duty diode that uses high current to convert AC to DC. |
| Registered Jack Connector | A type of modular connector that is made of "plug-in" units consisting of multiple, parallel contacts that can be added together to increase the size or capabilities of the system. |
| Register | A collection of Flip Flops that moves stored bits in one or both directions, usually serially. |
| Regulator | A device or circuit that maintains a desired output under changing conditions. |
| Relay | An electromechanical device that uses a small voltage to actuate a "switch." |

| | |
|---|---|
| Reprogrammable PLD | Each fuse of an OTP device is replaced with a static RAM, or SRAM, cell that allows the device to be reprogrammed while remaining on the circuit board. |
| Reset Line | Allows the counter to reset and restart an execution. |
| Resistor | A device that restricts current flow. |
| Resonance | Occurs at a particular frequency causing electrical energy to oscillate between the magnetic field of the inductor and the electric field of the capacitor. |
| RF Modulator | This modulator is employed to encode sine waves. An antenna is used to broadcast the sine waves as radio signals. |
| RF Oscillator | A device used to generate frequency and categorized as either an electronic circuit or a crystal oscillator. |
| RF Receiver | An electronic device that separates all incoming signals from each other and converts these signals into the desired format: audio, video, or data. |
| RF Transmitter | An electronic device that creates varying electric currents, encodes sine waves, and broadcasts radio waves. |
| RJ Connector | See Registered Jack Connector. |
| Rocker Switch | A switch that has an actuator that, when actuated, resembles the movement of a rocking chair, hence the name. |
| Rotary Encoders | An encoder that consists of a rotating disk, a light source, and light detector. |
| Rotary Potentiometer | A rotating knob or shaft used to control the amount of current that flows through a circuit. |
| Runaway Software | Software that has gone out of control or is now unstable. |
| Schottky Diode | A semiconductor diode (named after German physicist Walter H. Schottky) with a low forward voltage drop. |
| Second Generation Products | Products that are analog and digital current networks operating on 800MHz or 1900 MHz spectrums. |
| Sensor | A device that detects a single type of data and converts it into another form. |
| Signal Compression | Refers to reducing the number of bits required to describe a signal to a prescribed accuracy. |
| Signal Conditioning | Refers to the manipulation of electronic data so it can be more easily read by humans or other electronic devices. |
| Signal Level | The signal power or intensity at a specified point with respect to a specified reference level. |
| Single-phase Bridge Rectifier | See Diode Bridge. |
| Single-turn Potentiometer | A potentiometer that is used as an angle position sensor or as a panel controller when the rotation is less than a full revolution. |

| | |
|---|---|
| Slew Rate | The inability of the amplifier output (in operational amplifiers) to keep up with rapid changes in input. |
| Slide Potentiometers | A potentiometer that serves the same purpose as a linear potentiometer; it is adjusted by sliding a knob up and down a rail. |
| Slide Switch | A switch that is named after the sliding actuator; used to bring the switch to its actuated position or positions. |
| Snap Action Switch | A mechanical switch that produces a very rapid transfer of contacts from one position to another. |
| Solar Cell | An optoelectronic device with a large-area PN junction in a semiconductor. |
| Solid Electrolytic Capacitor | See Dry Electrolytic Capacitor. |
| Solid-State Relay | A complex semiconductor device that has no moving parts. |
| SPLD | Single Programmable Device |
| SRAM | See Static Random Access Memory. |
| SSR | See Solid-State Relay. |
| Static Random Access Memory | A type of random access memory that stores each bit of data in a separate capacitor. |
| Switch | A simple device that allows electric current to flow when closed and prevents current from flowing when opened. |
| Synchronous Counter | A type of counter in which the same pulse clocks each stage. |
| Tact Switch | A switch that is designed to have a small actuating force. |
| TAM | See Total Available Market. |
| Tantalum Capacitor | A capacitor with dielectric made of tantalum. This alloy is used because of its high melting point, high strength, and good ductility. |
| TCXO | See Temperature Compensated Crystal Oscillator. |
| TDM | Time Division Multiplexing |
| TDMA | See Time Division Multiple Access. |
| Telecom Connector | See Registered Jack Connector. |
| Telephone Connector | See Registered Jack Connector. |
| Temperature Compensated Crystal Oscillator | Similar to the Crystal Oscillator, but the temperature characteristic is corrected by the addition of a temperature compensation (thermistor) network. |
| Thermoelectric Cooler | A solid-state device that uses a thermoelectric effect and semiconductor materials to function. |
| Third Generation Products | Standards that enhance services, such as more extensive roaming capabilities and higher transfer rates. |
| Threshold | The point at which a signal (voltage, current, etc.) is perceived as valid. |

| | |
|---|---|
| Thyristor | A solid-state semiconductor device similar to a diode with an extra terminal. |
| Thyristor Gate | Used to turn on a thyristor. |
| Time Division Multiple Access | A form of WWAN that delivers digital wireless service using TDM. |
| Timer | A specialized type of clock that is used to control a sequence of events or a process.<br>See Discrete Logic Timer. |
| TOF Timer | Timer Off Delay |
| Toggle Switch | Actuated to open or close the circuit and toggle handles; can have various lengths, colors, and options, such as LED-lit. |
| Tolerance | The plus/minus percentage value that determines the accuracy of a resistor. |
| TON Timer | Timer On Delay |
| Total Available Market | The total world-wide revenues for the focal product. |
| Transceiver | See Data Interface Transceiver. |
| Transient Voltage Suppressor | Designed to absorb a transient over-voltage. |
| Transistor | A small electrical device that can either amplify an electrical signal (modulate) or switch it on and off, letting current flow or blocking it. |
| Transmission | The transfer of data over a communications channel. |
| Transmitter | See RF Transmitter.<br>See Data Interface Transmitter. |
| Trimmer | A Potentiometer that is similar to a rotary pot used in trimming the resistance in electronic circuits and usually mounted on a printed circuit board (PCB). |
| Trimming Potentiometer | See Trimmer. |
| Tunnel Diode | A diode with a negative resistance region. The current through the diode increases up to a certain voltage, than with increasing voltage, the current decreases until a voltage point is reached, and the current begins to increase again. |
| TVS | See Transient Voltage Suppressor. |
| UMTS | See Universal Mobile Telecommunications Systems.<br>See Third Generation Products. |
| Universal Mobile Telecommunications Systems | See Third Generation Products. |
| Universal Serial Bus | Emerging standard for interconnecting PCs with peripherals. |
| USB | See Universal Serial Bus. |

| | |
|---|---|
| Varactor Diode | A type of diode principally used as a voltage-controlled capacitor; its diode function is secondary. |
| Variable Resistor | A resistor whose value can be adjusted by a mechanical movement. |
| Variable Capacitor | A capacitor with a capacitance value that can be changed or set by some adjustment mechanism. |
| Varistor | A type of resistor with a non-ohmic current-voltage characteristic. |
| Voice-over Internet Protocol | The transmission of voice traffic over IP-based networks. |
| VoIP | See Voice-over Internet Protocol. |
| Volatile Memory | A type of memory that loses its contents when the power is turned off. |
| Voltage Controlled Crystal Oscillator | A Quartz Crystal Controlled Oscillator. When an external voltage is applied to the control pin, the frequency changes in direct proportion to the voltage applied. |
| Voltage Divider | See Potentiometer. |
| Wavelength-division Multiplexing | Sending several signals through one fiber with different wavelengths of light. |
| WCDMA | See Wideband Code-Division, Multiple-Access. |
| WDM | See Wavelength-division Multiplexing. |
| Welding Diode | A diode with a very large forward current rating. |
| Wet Electrolytic Capacitor | A type of aluminum capacitor that has a moist electrolyte. |
| Wideband Code-Division, Multiple-Access | A third-generation technology that uses the CDMA air interface rather than TDMA in GSM systems. |
| Wi-Fi | Wireless Fidelity<br>See Wireless Local Area Network. |
| Wireless Hotspot | A place for users to have an access point to a wireless network. |
| Wireless Local Area Network | Allows for the users in a local area to gain network access. |
| Wireless Local Area Network | A local area network that uses radio waves instead of wires or cables to communicate between its nodes. |
| Wireless Metropolitan Area Network | Allows for multiple networks in a metropolitan area to connect. |
| Wireless Personal Area Network | Primarily made up of Bluetooth technology. Allows users to connect personal devices within 30 feet of the set-up. |
| Wireless Wide Area Network | A wireless network that spans over large areas, such as cities or countries; uses satellite systems and antenna sites to make connections. |
| WLAN | See Wireless Local Area Network. |
| WMAN | See Wireless Metropolitan Area Network. |

| | |
|---|---|
| WPAN | See Wireless Personal Area Network and Bluetooth. |
| WR Line | See Write Line. |
| Write Line | Copies data from the computer main memory to a storage device, such as a disk or flash drive. |
| WWAN | See Wireless Wide Area Network. |
| XO Oscillator | See Clock Oscillator. |
| Zener Diode | A diode designed to conduct in the reverse [bias] direction, which means it will not let current flow until it has reached the breakdown voltage. |

# Index

## 2



## 3



## A

## B



## C



## D

# E



# F

## G



## H



## I



## J

## N



## O



## P

## Q



## R



## S

## T



## U



## V